Got Sun?
Go Solar

EXPANDED 2ND EDITION

Harness Nature's Free Energy
to Heat and Power Your Grid-Tied Home

Rex A. Ewing *and* Doug Pratt

PixyJack Press INC

Got Sun? Go Solar, Expanded 2nd Edition:
Harness Nature's Free Energy to Heat and Power Your Grid-Tied Home

Copyright © 2009 by Rex A. Ewing and Doug Pratt

Published by PixyJack Press, Inc.   PO Box 149, Masonville, CO 80541 USA

SECOND EDITION 2009
FIRST EDITION 2005

9 8 7 6 5 4 3   2   1

ISBN-13 digit: 978-0-9773724-6-1
ISBN-10 digit: 0-9773724-6-4

LIBRARY OF CONGRESS CATALOGING-IN-PUBLICATION DATA
Ewing, Rex A.
   Got sun? go solar : harness nature's free energy to heat and power your grid-tied home
/ Rex A. Ewing and Doug Pratt. -- Expanded 2nd ed.
       p. cm.
   Includes index.
   Summary: "Examines renewable energy options for grid-tied homeowners, including
solar- and wind-generated electricity, solar water heating, passive solar, and geothermal
heating / cooling. System configurations and equipment, average costs, financial incen-
tives, and installation considerations are also covered"--Provided by publisher.
   ISBN 978-0-9773724-6-1
1. Photovoltaic power systems. 2. Solar energy. 3. Wind power. 4. Wind turbines.
5. Dwellings--Electric equipment. I. Title.
   TK1087.E95 2009
   697'.78--dc22

                                    2009019053

Printed in Canada on chlorine-free, 100% recycled paper (post-consumer waste).
By printing on 100% recycled paper, we save 42 full-grown trees, 3,408 pounds of greenhouse
gases, 15,415 gallons of water, and 1,798 pounds of solid waste — for every 5,000 books.
SOURCES: www.greenpressinitiative.org • www.environmentaldefense.org

Book design by LaVonne Ewing.

DISCLAIMER: Due to the variability of local conditions, materials, skills, building site and so forth, PixyJack Press, Inc. and the authors assume no responsibility for personal injury, property damage, or loss from actions inspired by information in this book. Electrical wiring is an inherently dangerous activity, and should be regarded as such. Always consult with your local building codes and the National Electric Code before installing and operating renewable energy systems. When in doubt, ask for advice; recommendations in this book are no substitute for the directives of equipment manufacturers and federal, state and local regulatory agencies.

To LaVonne Ann,
Forever the sunshine in my soul
and the wind at my back.

– Rex

To my children and
their children's children,
in the hope that mankind will
learn to thrive within its means.

– Doug

Contents

continued

continued

Section Two: Heating Your Home with Nature's Free Energy

Introduction

Have you ever gotten so tired of waiting for someone to do something for you that you finally just did it yourself? That's how most of us learn to fix our bicycles when we're kids: we get fed-up waiting for some disinterested adult—who will probably want an extortionary favor in return—to grudgingly agree to adjust the brakes or patch a flat tire. Exasperated, you march off to the garage, fish around for a few tools, and get to work on the problem. By the time the repair is done and you're glowing with a self-satisfied sense of accomplishment, you realize it really wasn't all that difficult.

Sound familiar? Then this expanded 2nd edition is for you. The air these days is buzzing with talk of our expanding national commitment to renewable energy, and we applaud the government's efforts that set a more sustainable course for our country's energy future. But the only way to personally be an active part of this green energy is to make a resolution to wean your house or condo (or, yes, even your apartment) from the fossil-fuel cow. It's not as hard as you may think: with a bit of planning and a few components, you can convert your little corner of the world into an environmentally responsible haven you can be duly proud of.

Today's photovoltaic (solar-electric) panels are cheaper, more efficient and versatile than they've ever been, and when combined with modern power inverters they will generate electricity that is "clean" enough to satisfy the persnickety needs of the most delicate electronics. Whether you want a large array to run an entire house or one PV panel for your apartment, solar electricity can be configured to fit your needs.

Got wind? The noisy, awkward, breakdown-prone wind turbines of the past have been replaced with quiet, highly efficient marvels of engineering that can produce useful amounts of power in a gentle breeze. Many of these new-generation wind turbines can be directly tied into the power grid without the need of batteries or charge controllers.

This is all really cool stuff that wasn't available a few years ago, and it's just waiting to be installed on or near your home.

Do you like the idea of legally spinning your electric meter backwards, and doing it with a simple, non-polluting, silent power source that will outlive your children? That's what solar energy can deliver right now, and this book will explain your options. Solar-electric and wind systems deliver their energy directly to your household, with any surplus pushing out through your meter into the grid. This is called utility intertie, or simply grid-tie, and it's legal in every state.

But solar- and wind-electric systems are just the beginning. There is also solar hot water to consider. It is, after all, an idea that's been in use since Roman times. And if you watch with dismay as the rising price of natural gas or electricity turn your morning shower into an increasingly expensive proposition, you may want to consider a solar hot water system for your domestic hot water needs. There's a system for practically every climate and budget, and unlike the showy, clunky installations of the Carter era, recent innovations in heat-collection technology combine efficiency with class and discreetness. Best of all, state and federal incentives for solar hot water systems are also there for the taking.

But not all renewable energy comes directly from the sky. Geothermal energy is perhaps the most efficient and least under-stood form of free and inexhaustible energy. Modern ground-source heat-pump systems are designed to extract the heat from the ground beneath your feet so quietly and efficiently you'll feel like you're pilfering heat on the sly. It's the perfect hedge against rising heating-fuel prices.

So what are you waiting for? Whether you're a trained electrician or plumber qualified to do the work yourself, or just a home-owner who believes that waiting until tomorrow to make the world a better place is not an option, you can use the information in this book to help you decide which system or systems will work best for you. We'll even give you tips on making the most of passive-solar home design and also how to conserve energy in your home to achieve the greatest impact from nature's free and renewable energy sources. By the time you get around to finding an installer, you'll be on intimate terms with your future system.

If you still need a little push, the federal government and most states offer financial incentives for those willing to invest in solar, wind and geothermal energy, including tax credits and substantial rebates. And should you need financial assistance to get the process kick-started, we'll tell you how to go down that road, too.

Let's face it: you are all out of excuses.

Renewable energy is a proven technology that is affordable and adaptable to almost any situation, and the financial incentives are about as sweet as they can be. Whether you decide to travel down the path to renewable energy on an off-the-shelf Schwinn or a custom-built Madone SSL by Trek, it's going to be your bicycle, so make the best of it. The wait is finally over. ❖

The First Modern Solar Cell

It wasn't exactly serendipity, which is, as they say, looking for a needle in a haystack and finding instead the farmer's daughter. No; back in 1952, Daryl Chapin from Bell Labs was definitely looking for the farmer's daughter from the get-go. He just didn't find her in a haystack.

Here's how it happened: Researchers at Bell labs, headed by Chapin, were trying to find a way to operate Bell telephones in remote places, of which there was no dearth back then. Dry cell batteries were short-lived, especially in hot, humid climes, so Chapin was searching for a more satisfactory source of power. Wind was considered, along with thermo-electric power and even steam engines. Chapin, being a solar enthusiast (there was a great deal of interest in passive solar after World War II, due to a worldwide fuel shortage), sug-

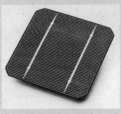

Crystalline silicon solar cell.

gested that the idea of photovoltaics be explored.

At that time, the only PV (solar) cells in existence were made of selenium. Selenium cells, however, could only produce about 5 watts per square meter; a mere 0.5% conversion efficiency. Chapin wanted 6.0%—a twelve-fold increase in efficiency.

As the gods of research would have it play out, two of Chapin's colleagues, Gerald Pearson and Calvin Fuller, were working in a nearby lab with crystalline silicon in hopes of building a solid-state rectifier, a device that transforms AC to DC. But pure silicon is a not a very good conductor of electricity. Fuller, however, managed to improve its conductivity appreciably by introducing gallium into the crystal matrix. Then Pearson took it a step further; he gave the gallium-rich silicon a bath in hot lithium.

For whatever reason—scientists do so love to play—Pearson shone a light on his crystal and discovered that the light energy induced an electric current. (He certainly must have shouted something like "Eureka!" at that point, though his exact words are lost to history.) Pearson rushed to Chapin's office to announce the good news.

Though the first, crude silicon solar cell was not capable of the 6% efficiency Chapin was aiming for, it was far better than anything he'd yet found. Chapin went to work to improve it. The big breakthrough came when Fuller vaporized phosphorus onto the surface of a nascent solar cell. This brought the p-n junction (of which you will read more, shortly) near the surface and, perhaps more importantly, allowed for the deposition of conducting channels to carry the light-induced current away from the cell and into an electrical circuit.

By 1954, Chapin, Pearson and Fuller had produced a 6%-efficient solar cell and announced it to the world. Much of this book follows from their discoveries…

SOURCES: *Solar Today* (Jan/Feb 2004) "Good as Gold: The Silicon Solar Cell Turns 50" by John Perlin, Lawrence Kazmerski, Ph.D., and Susan Moon; and *www.eere.energy.gov/solar/solar_timeline.html*.

PV panels power the space station Skylab (above) and an emergency radio tower in remote mountains (right). *PHOTOS: NASA; SHELL SOLAR*

In the late 1950s, solar cells were first used in space, and they quickly became the most widely accepted energy source for space applications. In 1964 NASA launched the first Nimbus spacecraft – a satellite powered by a 470-watt PV array. Since then, numerous craft have been launched, including Skylab and the Mars Rover. Solar cells are used today to provide power for remote telecommunications, signals and sensors, navigation aids, water pumping, highway call boxes, off-grid lighting, calculators,

watches, portable electronics, medical clinics, cathodic protection to prevent iron corrosion, security lighting, billboards, and emergency highway signage… not to mention our homes and offices.

Generating Electricity from the Sun and Wind

Solar Electricity

Wind Power

CHAPTER ONE

Why Would You Want A Solar/Wind Electric System?

There are many good reasons to install alternative energy, some of them green, fuzzy, and warm; some of them hard economic facts. Your authors are a cowboy storyteller and a big soft techie so we'll cover the warm fuzzy stuff first because it's more fun.

Warm, Fuzzy Reasons for Renewable Energy

Solar energy isn't diminished by harvesting. The amount of energy we take today in no way diminishes how much we can take tomorrow, or how much is left for our children or grandchildren. Every single day enough solar energy falls on the Earth to supply all the world's energy needs for four to five years. Solar energy shows up directly as sunlight, which is harvested by panels that either create heat or electricity. Our allotment of solar energy can also show up indirectly as wind—the result of uneven heating on the Earth's surface. Large commercial-scale wind harvesting is a well-developed, rapidly growing industry, and residential-scale wind is seeing some nice advancements. Solar energy also appears as rain. If precipitation falls at higher elevations we can harvest great amounts of energy from falling water—hydro power—another subject that needs its own book. From our allotment of incoming solar energy, the incredibly tiny fraction of

Everyone's Going Solar
Over 5,948 megawatts of solar electric modules were installed worldwide in 2008. That's up from 547 megawatts in 2003, and light-years ahead of 21 megawatts in 1983.

a percent that we could potentially harvest will have no effect on world weather. But it can have a dramatic effect on your own well-being, and on the well-being of your offspring, by reducing demands on finite fossil fuel reserves, and by reducing the amount of carbon dioxide (CO_2) that is released to supply your power needs. By supplying some, or all, of our energy needs directly from sunlight we leave more resources for the future, we reduce global warming, and we become better world neighbors.

The kind of panels that turn sunlight into electricity are called photovoltaic modules (say *photo-voll-ta-ick*), or PV modules for short. PV modules have no moving parts, they are silent and non-polluting. They have manufacturer warranties of 25 years, and useful life spans that will probably exceed 60 years. The embodied energy input required to manufacture them is repaid in two to four years according to studies from the National Renewable Energy Labs (to read the condensed, two-page report: *www.nrel.gov/docs/fy04osti/35489.pdf*).

Not that we really need to mention it, but some folks will feel a considerable sense of relief knowing that none of a PV module's raw materials come from the Middle East. Photovoltaic cells—the dark parts of the module—are made from high-grade silica, the same raw material in computer chips. It's a primary component of beach sand, although most silica today comes from mining operations, often as a byproduct. The PV modules use glass covers (more silica) and are highly recyclable. The frames of most residential modules are aluminum. That's where all those beer cans go once they've finished their wild youth and need to do something useful with the rest of their lives. In addition there's a small amount of plastic, used for sealing back covers and junction boxes. Some of it is recycled already; all of it is recyclable in that long-distant future when a PV module has finished its useful life. The mounting structures that support the modules are almost always aluminum with stainless steel hardware.

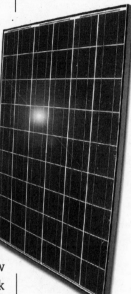

Even after subtracting the CO_2 produced during manufacture and mounting operations, this 187-watt module can be expected to produce enough electricity during the next 30 years to offset the emissions of over 13,600 pounds of CO_2 from utility power.
PHOTO: KYOCERA SOLAR

Calling a Spade, a Spade

Solar panels that make
electricity are called
"photovoltaic modules"
or PV modules for short.
This separates them
from solar collectors
that make hot water,
which is a very different
technology. Water and
electricity usually don't
mix, and that's certainly
true for PV power.
In this section, we'll be
talking about PV modules,
or just modules.

If there's any part of a modern PV module that started its industrial life under some Mideast sands, it's a trifling fraction of a percent that could just as well come from someplace less volatile. There's a real lack of guilt or exploitation wrapped up with PV modules. If that really bothers you, I suppose you could go rent a Hummer for a weekend to balance your karma.

Hard-Edged Financial Reasons for Solar/Wind Power

You can get a real warm, fuzzy, and highly self-righteous feeling from using renewable energy (RE), but what's it do for your wallet and well-being? A solar/wind electric system allows you to cover part, or all, of your electrical needs yourself. Every watt-hour your system delivers is a watt-hour you don't have to buy from your utility company. Once installed and connected to the utility grid, your RE system

Enphase micro-inverters are perfect when needing to install solar panels on two faces of the roof, as shown here. *PHOTO: ENPHASE ENERGY*

will reliably shave your most expensive kilowatt-hours off the top of your bill every month. It offers shelter and protection from rising utility rates in the future. Your solar system is going to last as long as your house, and it's going to reliably crank out watt-hours every day. Your house can have half the electric bill of your neighbors for its entire life! What's that worth? If needed, your solar electric system can be configured to continue providing power to selected circuits during utility blackouts (see chapter 4). This costs a little more, but in storm-battered locales, what are continually operating furnace, fridge, and lights worth to you?

The federal government and many states and utility companies are offering rebates, tax credits, or other incentives for installing solar and wind power. Some of these programs will cover more than 50% of the installed cost, which can bring simple payback periods

A HUGE Array

A 100-mile x 100-mile PV array in Nevada would provide all the electric power for the United States.

Say "Thanks!" to Northern California Pot Growers
(and not necessarily for their commercial product)

In the early days of photovoltaics – the 1980s – a curious thing happened. Northern California was overrun with back-to-the-land hippies who'd found cheap, beautiful land and an agreeable climate back in the hills of Sonoma, Mendocino, and Humboldt counties. Much to their surprise however, they found that money didn't grow on trees. But it did grow on bushes! At least on some kinds of bushes. So they did what came naturally, they became growers. This cash crop allowed them to buy groceries, used 4-wheel-drive Toyotas, building materials, and eventually niceties like electric lights. Being miles from the nearest power line, they bought RV lights they could run off a battery. Hauling the battery to town or running the generator to recharge it got old pretty quick, so when the very first solar electric panels started showing up, they embraced the technology wholeheartedly, despite PV costs that were many times more expensive than now. All through the 1980s more than half of the entire U.S. sales of PV modules ended up someplace between Santa Rosa and the Oregon border. It was growers who could afford them, growers who wanted them, and to a very large degree, growers who financed the early development of the PV industry.

Another Step Forward

Starting January 2007, California's SB1 state law provides $3.2 billion in funding for solar roofs over the next ten years.

well under 10 years. Most folks will see longer payback periods, especially if you assume utility rates won't rise (yeah right). But every solar electric system will eventually repay its installed cost, probably long before the PV warranty expires, and certainly decades before they wear out. Clean Power Research has developed several software tools that provide highly objective economic analysis of investments in clean energy technologies. Their web-based estimator, with preloaded climate, utility rates, and available rebates, is posted by several states with rebate programs, including California, New Jersey, Florida, Ohio, Hawaii, and probably others by the time you read this. Simply do a Google search for "Clean Power Estimator." ❖

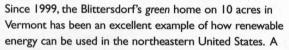

Since 1999, the Blittersdorf's *green* home on 10 acres in Vermont has been an excellent example of how renewable energy can be used in the northeastern United States. A 10 kW Bergey wind turbine and 7.0 kW solar array provide all of their electricity, and under a Vermont Public Service Board permit "Certificate of Public Good," they are net metered. The top photo shows a sunroom with PV panels on the roof and the wind turbine. In the lower right corner of the right photo, you can see more of their PV array. Solar hot water panels are also used to heat the swimming pool and home. *PHOTOS: DAVID BLITTERSDORF*

CHAPTER TWO

Is This Stuff Legal...or Safe?

Way back in 1978, mostly as a result of the '73 Arab oil embargo, and to encourage renewable energy development, the federal government passed the Public Utilities Regulatory Policies Act (PURPA) which says that any private renewable energy producer in the USA has the right to sell excess renewable energy to their local utility company. Until PURPA, only large-scale mills and factories were able to sell excess energy back to the grid. Now the door was open to everyone. The federal law didn't say that utilities had to make this easy or profitable, however. And believe me, it wasn't/isn't either. Interconnection standards were whatever your local utility said they were, which made it nearly impossible for small-scale wind and solar manufacturers to develop and mass-produce standardized equipment. And without standards, the utilities justifiably worried about safety. Would this customer's equipment shut off automatically if utility power went off? Demands for the customer to carry million-dollar liability insurance were not uncommon. Payment wasn't anything to get excited about either. Under PURPA rules, the utility has to pay their "avoided cost," which is usually defined as their wholesale rates, but can get sticky because wholesale rates vary widely depending on time of day and

PV: A Growth Industry

Solar energy demand has averaged a 25% yearly growth rate over the past 15 years, with the pace accelerating since 2005. While conventional utility grid energy demand averages under 2% yearly growth.

season. Usually it works out to around two to four cents per kilowatt-hour. None of this was the least bit encouraging, and renewable energy development in the U.S. languished throughout the 1980s.

Net Metering

In the mid 1990s individual states started passing "net metering" laws which allowed small-scale producers to very simply sell excess renewable energy through their existing meter to the utility for standard retail rates. This means you could push a kilowatt-hour of energy into the grid during a sunny afternoon, buy a kilowatt-hour back later that evening, and enjoy a total of zero on your meter. This sure beats selling it for 2 cents, then buying it back for 10 cents. Net metering laws are a very good deal for renewable energy producers because it allows us to treat the utility like a big, never-full, never-wears-out, 100%-efficient battery that we don't have to buy, service, or maintain. It isn't such a great deal for utilities, who get to buy your kilowatt-hour for 10 cents, then sell it to your neighbor for 10 cents. For this reason, most utilities have negotiated caps on how much net-metered power they're required to purchase, usually expressed as a percentage of their total, and how long a customer can hang onto a credit.

A home on Long Island, New York with a 2.7 kW array of Evergreen solar modules. *PHOTO: EVERGREEN SOLAR*

As of mid-2009, there is not a federal net metering law, but forty states have statewide laws, and several states have individual utilities that actively support net metering. To see how net metering is becoming the norm, and to check out your state, see *www.dsireusa.org*.

What if you're not in a net metering state? All is not lost, but first contact your state representatives and ask why not...why is your state in the minority? Get on the ball! Then contact your friendly local electric utility and ask what their policy is with small-scale renewable energy producers. We've heard stories from customers about small utility companies that, when faced with the choice of developing a unique billing system for a single customer, or even a handful of them, simply take the easy way out and allow net metering. If that doesn't fly, you might consider a smaller system that doesn't sell back much. Remember that every electron your solar/wind electric system delivers is one electron you don't have to buy. You're getting the same one-to-one effect as net metering, up until your production starts to exceed your household use.

Earning Credit at Your Utility Company

Net metering means you can "sell" your extra kilowatts back to your utility company, if they allow it.

Do I Need a Special Electric Meter?

In most cases the answer is no, but it's probably going to be a different meter than the one you've got now. The common mechanical clock-work meters will run forward or backward just fine. They aren't, however, calibrated to record accurately in reverse. Utilities usually provide a calibrated single-rate, bi-directional meter at little or no charge...after all, it's their meter. If you're on a time-of-use billing scheme, then you'll need a very special meter, and that may have a one-time installation charge.

Occasionally a ratchet is installed on a meter so it can only run in one direction. This is often how differential buying and selling rates are handled. Two meters are installed in series. One will only read input, the other will only read output.

The new digital meters are another story. Many of the fully digital meters simply have a little magnet on the wheel. It counts one revolution every time the magnet goes by, assuming that it only goes by in the forward direction. You could be paying for every kilowatt-hour you send out! I actually caught my helpful utility company installing a meter like this (twice!), when they thought they were installing a bi-directional meter. It had a "bi-directional" sticker inside, but it didn't register bi-direction-ally, and managed to get on two different service trucks on two different days. So always check — you want to see that baby registering backwards!

— *Doug Pratt*

Rebates and Incentives

A great resource for
the latest information on
rebates and incentives
is the **Database of
State Incentives for
Renewable Energy**
at *www.dsireusa.org*.
Just click on your state
for a complete listing of
rebates, incentives, tax
breaks, or any other way
that money is available
for clean energy
in your state.

Why Your Utility Company Doesn't Want Your Power

Utility companies are naturally going to be very concerned about any source that could be feeding energy into their network. Is the frequency, voltage, and waveform within acceptable limits? The utility is going to accept your power only if it's as good or better than the product they normally deliver. Is it clean enough to sell to the neighbors? The utility is responsible for any power quality problems once they accept your solar-generated electrons. What happens if the utility power fails? Line workers could be at risk if your system continues feeding power to the grid after a failure. There are many highly valid reasons for utility companies to be less than enthusiastic about accepting power input from small independent producers. Once it's in their system, they bear total responsibility if something happens to the neighbor's TV due to power quality issues.

Technology Leads...

Trace Engineering started producing their intertie-capable SW-series inverters in about 1994–95. This was the first really affordable intertie inverter. It brought the possibility of utility intertie within reach for everyday folks. And it was so easy! Just connect your PV modules to your batteries like usual, connect your batteries to your inverter like usual, and connect your inverter AC1 terminals to utility power, unlike the usual. (The AC2 terminals went to the usual generator.) Folks started hooking them up to utilities and pushing the "sell" button almost immediately...sometimes with official approval, but often without it. It worked fine either way. Funny how software doesn't care about legalities.

Meanwhile, various states were passing laws that said if their citizens wanted to sell excess renewable energy back to the grid, the utility

had to accept it. And in many cases, they had to accept it via net metering, which gave retail rate credits to the homeowner. Utilities were scrambling. On the one hand, state law said they had to accept this power. On the other hand, they had no idea how clean, safe, or regulated this power was. There were several years where utilities, in order to pro-
tect themselves, had to go buy examples of grid-tie (intertie) inverters, set them up, and test them for themselves. PG&E of California, the nation's largest utility company, had a list of accepted inverter models. A very short list. So long as you installed something on their list, you could get approved for net-metering intertie. Many smaller utility companies either stonewalled, because they couldn't or wouldn't deal with the confusion, or simply accepted anything on PG&E's list.

A very large residential array that averages about 50 kWh per day.
PHOTO: SHELL SOLAR

...And Regulations Follow

Meanwhile, in the background, standards and regulations were being developed as quickly as possible. UL standard 1741 was first unveiled in May 1999 to address "distributed generation," a new class of electrical equipment that can potentially push power back into the grid. Here was the small-scale interconnection standard we needed, but never had, all through the 1980s and '90s. The UL1741 interconnection standards have already been updated and expanded several times, as is normal for new standards. It sets precise limits for how low or high the voltage can wander, how much distortion is allowed in the waveform, what has to happen under any conceivable utility condition, and how fast it all has to happen. As it has been updated, UL1741 has

UL1741 Gets A Makeover
In May 2007, a major upgrade to UL1741 took effect. All grid-tie inverters manufactured after this date offer improved grounding, neutral-sensing, and other safety improvements.

On average, for every $1 in incentive committed by the California Solar Initiative, an additional $6 in private funds is invested in solar technology in California.

SOURCE: CALIFORNIA PUBLIC UTILITIES COMMISSION

been incorporating industry standards for distributed generation power delivery as they've been developed. In addition to UL1741 compliance, some inverter manufacturers will list a series of IEEE (Institute of Electrical and Electronics Engineers) standards. IEEE 519, which defines power quality standards, and IEEE 929 which defines anti-islanding, are the most common. The end result is that the output power from a grid-tie inverter is held to much higher standards than your local utility has to meet. UL1741 guarantees better-quality power to the homeowner than your local utility can promise. All grid-tie inverters on the market are now certified to this standard, which means they have been rigorously (expen$ively!) tested and proven to deliver cleaner, tighter, better regulated power under all conditions. If your utility power fails, or the power quality wanders outside the standards, the inverter will stop selling power within a

Turning On Northern California's First Residential Grid-Tie System

Back in late 1995, I had the dubious pleasure of punching the "sell" button on what was apparently PG&E's first residential solar electric intertie system with net metering. At least it was the first one that PG&E knew about, and the pleasure was dubious because...well, let me tell you the story.

In '92 and '93 I had designed and consulted on the installation of a nice hybrid solar electric/wind system at the north end of the Napa Valley. Even though the customer (we'll call him George) was right next to the county road with readily accessible power lines, he chose to build his passive solar house off-grid. This was intended as George's retirement home, and no power bill would be one less future uncertainty. But retirement was several years away yet, so

for tax purposes he made it a rental property in the meantime.

For the most part this worked out fine, except for a few times in the depths of winter when there was no sun for two or three weeks. Neither the tenants nor George liked dealing with generators. So when the State of California announced that as of January 1995 small grid-tie systems would be legal, George was first in line with his application. This was a strange new critter to PG&E, and they took their time granting approval. In the meantime we did a little equipment upgrading and George was ready by March. It was late September when the local electrician finally called to tell me George had been approved for turn-on in early October. And, oh, by the way, a couple maximum of 75 milliseconds. Most utilities agree this is probably quicker than their repair crews on most days.

Even exceedingly unlikely scenarios are anticipated, like islanding. A utility island happens (theoretically) when the utility power fails, but leaves a small isolated area like you and a few neighbors who happen to be using precisely as much energy as your grid-tie system is delivering at the moment. So long as everything stays in balance (a highly improbable assumption) such an island could keep running for a long time, except that some over-caffeinated official already thought of it. So every couple minutes the inverter will do a little hiccup as an island check. The point is, UL1741 is exhaustively thorough. Every possible problem, and even a few impossible ones, have been considered, and if necessary, a performance standard has been written to protect all involved.

PG&E bigwigs wanted to be there when we officially punched the sell button. Would I mind being on hand to answer any questions? No problem. The site was an hour and a half of very scenic twisty roads from my work. It was a great ride on my beemer.

So on the appointed day I showed up, dressed in what I thought was appropriate garb (red and black motorcycle leathers). Much to my surprise, PG&E had brought an entire bus! And it was mostly full of engineers, each of whom had at least twenty urgent questions! I was quickly surrounded and interrogated. Now I know how celebrities feel. Fortunately for me, I'd packed several extra copies of the copious Trace SW-Series Owner's Manual that I passed out to the crowd. It was like tossing raw meat to a pack of wolves; they all went off to corners to devour the goodies. I knew from talking to the Trace folks that PG&E had been testing sev-

eral of these inverters in their labs for over six months. I guess their engineers are fairly spread out and don't get to talk with each other as much as they should.

The time came to activate George's system. I went into the power room, while most of the PG&E crew went outside to watch the meter. I punched the "sell" button, enabling the inverter to start selling excess PV power. Outside, as the meter dial slowed, stopped, and then started rotating backwards I could hear everyone go, "Oooohh!" After all, nobody had ever seen this before. This was quickly followed by shouts of, "Oh! Oh! Do it again! Do it again!" So we simulated a couple of PG&E power failures and demonstrated how it wouldn't sell if it couldn't sense utility power. This made everyone a lot more comfortable, and everybody went home pleased with the day.

— *Doug Pratt*

Is it Grid-Tie or Intertie?

These two words
describe the practice of
treating the utility grid
like a big battery.
Intertie is probably
more descriptive of the
give-and-take relationship,
but *grid-tie* will be
understood by more
folks. Once you own
one, you can call it
whatever you want.

Having a precisely written UL performance standard assures both utilities and customers that these inverters simply cannot present any threats to workers or owners. It has also helped curb most of the absurd insurance demands that some utility companies were attempting to impose on customers, such as million-dollar liability coverage. Most utilities have backed off, and several states have specifically outlawed any additional requirements for interconnection, so long as all components are UL listed. For several years this left an amusing artifact clause in the PG&E interconnection agreement, where a grid-tie customer was "...required to maintain no less liability insurance than they currently carry." Uh...sure. We can do that.

In addition to UL specifications, we're starting to see equipment with FCC (Federal Communications Commission) compliance ratings. An FCC rating isn't mandatory...yet, but it probably will be in a few years, and most manufacturers are voluntarily complying, especially as they develop new designs. The FCC Part 15, class B rating you're likely to find on your inverter says that this appliance will cause minimal interference with radios, TVs, and other common wireless devices. Note we said minimal, rather than no interference. Inverters all make some electronic noise. If you're close enough, with the right gizmo, you're still likely to get some interference. But your chances of being able to listen to the ball game on AM radio are greatly improved with this new generation of FCC compliant inverters. ❖

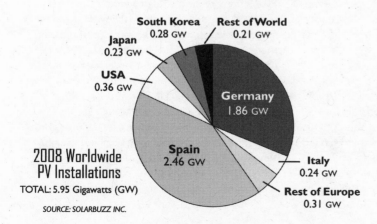

2008 Worldwide PV Installations
TOTAL: 5.95 Gigawatts (GW)

SOURCE: SOLARBUZZ INC.

South Korea 0.28 GW

Rest of World 0.21 GW

Japan 0.23 GW

USA 0.36 GW

Germany 1.86 GW

Spain 2.46 GW

Italy 0.24 GW

Rest of Europe 0.31 GW

CHAPTER THREE

Riding Herd on Electrons: Electricity and Solar Cell Basics

N ow that you are acquainted with the legalities of grid-tied solar and wind systems, it's time to get down to the *really* fun stuff...like, how does it all work?

Electrons Have All the Fun

When you're talking about things electrical, you are really talking about electrons and their highly predictable tendencies. Electrons, for those of you who have successfully managed to repress all memories of chemistry class, are the negatively charged particles that orbit atomic nuclei at breakneck speeds from comparatively great distances. They are unimaginably tiny things; so tiny, in fact, that it would take 166,000,000,000,000 of them laid out like beads on a string to span the period at the end of this sentence. Just the same, as diminutive as they may be, electrons hold everything together, since all molecules are collections of atoms eager to share each others electrons.

"Big deal," you say. "What's all this got to do with solar energy?" I was just getting to that.

When electrons aren't holding stuff together, they are usually involved in making electricity. Plug in a vacuum cleaner, turn on a radio, or stick a slice of bread in a toaster, and all you're really doing

Not Everyone Can Flip A Switch

Two billion people in the world have no access to electricity. For most of them, solar photovoltaics would be their cheapest electricity.
SOURCE: SOLARBUZZ.COM

Voltage is a measure of electrical potential (or pressure).

Amperage is a measure of the total amount of electricity that passes a given point over a given amount of time (a rate of flow).

Wattage is simply a measure of the amount of electrical power provided by the circuit (work performed).

is opening the floodgates to permit a multitude of electrons to flow from a place where they reside in abundance, to a place where there is a dearth of the evasive little things.

It's not a free ride, of course. We fully expect to get paid for our efforts. How do we collect? By putting obstacles in the electrons' path. Obstacles such as electric motors, induction coils, or heating elements. If electrons want to get from here to there—and they really, really do—they're going to have to do some work along the way. It's only fair; we just can't let them have all the fun.

Learning To Love Volts, Amps and Watts (but not Ohms)

Electricity, then, is simply the flow of electrons through a conductor from one place to another. Gazillions of them. And, if you hope to do some of your own system planning, there are exactly three terms you will need to acquaint yourself with: voltage (volts), amperage (amps), and wattage (watts).

Voltage is a measure of electrical potential. In electrical terminology, voltage is called the "electromotive force" of an electrical circuit. Though it's a shameful act of anthropomorphism, you could think of electrical potential as the desire a pack of electrons has to get from one place to another. The greater the desire, the higher the voltage. This is why a spark from the end of a spark plug wire will jump an inch or more to an engine block. At 30,000 volts, the electrons are highly motivated to get out of the wire and into the metal block to complete the circuit (though, if your hand is carelessly close to the end of the wire, the electrons will consider it a conductor, too).

Amperage, by contrast, is a measure of the total amount of electricity that passes a given point over a given amount of time. The rate of flow, in other words. Technically, an ampere is one volt of electricity passing through one ohm of resistance. (What's an ohm? It's a measure of resistance to the flow of electricity, and it will not figure into any of your calculations, unless you choose to go about them the hard way.)

That leaves wattage, which is simply a measure of the amount of electrical power provided by the circuit. A watt is defined as the amount of work done when one ampere at one volt flows through one ohm of resistance. (Really—you don't have to worry about ohms.) It is equal to 1/746th of a horsepower.

Putting these terms together, we learn that:

Volts x Amps = Watts

So, if your circular saw draws 5 amps of power when plugged into a 120-volt wall outlet, it will consume 5 x 120 = 600 watts of power. By reworking the equation, we can see that:

Watts ÷ Amps = Volts   and   **Watts ÷ Volts = Amps**

These three equations will likely be the basis for every bit of number crunching you'll need to do when designing and operating your system.

Rex's Raillery

One ampere is equal to 6,250,000,000,000,000,000 electrons passing a given point in one second. Running a 60-watt (0.5 amp) light bulb for one hour requires the services of 11.25 sextillion (1.125 x 10^{22}) electrons. This figure exceeds the number of special-interest lobbyists in Washington, D.C. — at least for the time being.

A patriotic Potomac, Maryland home sports a 3.6 kW array of SunPower PV panels.
PHOTO: NEVILLE WILLIAMS, STANDARD SOLAR

Electricity's Who's Who

The **volt** is named after Count Alessandro Giuseppe Antonio Anastasio Volta, the Italian-born physicist who invented the electric battery in 1800.

The **ampere** gets its name from André-Marie Ampère, a French contemporary of Volta who was the first to combine the phenomena of electricity and magnetism into a single electromagnetic theory.

Falling Water and Electron Flow: A (Nearly) Perfect Analogy

If you're still a little confused about the relationship between volts, amps, and watts, you are probably not alone. Fortunately, a little analogy should help bring things into focus.

Imagine that you have two tanks of water. One is 20 feet overhead, the other twice that high. Each has a hose attached to its outlet, sized to ensure that the amount of water flowing from either tank remains the same. Opening the nozzles on both hoses, you will notice that the hose attached to the loftier of the two tanks produces twice the pressure of the other—it will shoot a stream of water twice as far—though it will not fill a bucket any faster than the hose from the lower tank. That's because the water in the 40-foot tank has twice the potential energy (voltage) of the water in the 20-foot tank, but not a higher rate of flow (amperage).

How do we increase the amperage? Well, in this analogy, amperage is defined as the number of water molecules to pass a given point in a given amount of time, and it is independent of pressure (voltage); no matter how much you increase the pressure, with the original hose you can't increase the rate of flow. Therefore, the only safe way to raise the amperage is to increase the size of the hose. That's why thick wires can carry more current than thin ones. Of course, you could attach a pump to the end of the hose and pull water out of one of the tanks at a faster rate than the hose was meant to handle. This would be analogous to running power to a 30-amp welder with a skimpy lamp cord. Not a good idea.

This leaves us with wattage, the amount of work being done. To illustrate what wattage really is, we'll attach micro-hydro turbines to the ends of both hoses. As you might imagine, water flowing from the 40-foot tank will turn the turbine twice as fast as water from the 20-foot tank, thereby performing twice the work, even though the actual volume of water flowing through both hoses is the same. How do we get an equal amount of work from the water in the 20-foot tank? You guessed it—we double the carrying capacity of the hose. This is akin to doubling the amperage, and will give us the same wattage (amount of work done) as the smaller hose attached to the higher tank:

A 40-foot drop x 1 unit of carrying capacity =
a 20-foot drop x 2 units of carrying capacity

or: 40 volts x 1 amp = 20 volts x 2 amps

Isn't this simple? Don't get too complacent, however, because now I'm going to throw a little wrinkle into our nice, smooth analogy. It has to do with the inherent differences between rubber hoses and copper wires. For, whereas a rubber hose will eventually burst when subjected to enough pressure (voltage), a copper wire will suffer no such ill effects. The 30,000 volts coursing through a skinny sparkplug wire is a case in point. Because the amperage is negligible, the small wire can handle the current, even though the voltage is immensely higher than anything you would ever use in your home.

This non-intuitive fact leads to a practical conclusion: if you increase the voltage of your solar array, you will be able to use smaller wire to conduct the current to the house. In fact, by doubling the array voltage—from, say, 24 to 48 volts—you are quadrupling the carrying capacity of the wire. How? If you have a solar array capable of producing 960 watts of power, and if your system is 24 volts, you will need wire sized to carry 40 amps of current (960 ÷ 24 = 40). If you increase your array voltage to 48 volts, however, your wire only needs to be sized to carry 20 amps (960 ÷ 48 = 20). And, in addition to halving your amperage, you are also halving the percentage of the voltage lost due to resistance in the wire, hence you are increasing the wire's carrying capacity four-fold.

Repeat after me: Voltage is Good.

Electricity's Who's Who

The **watt**, on the other hand, is named after Scottish-born inventor, James Watt, whose passion was not electricity, but steam engines. Though he didn't dabble in electron flow, he did introduce the term "horsepower," and in 1889, 76 years after Watt's death, the Second Congress of the British Association for the Advancement of Science adopted the watt — equal to 1/746th of a horsepower — as the unit of electrical power.

Break a Deal,

Face the Wheel

(or the AC generator)

Nikola Tesla almost single-handedly invented AC power. Yet Tesla's first employer in America was, ironically, Mr. DC himself, Thomas Edison. When Edison stiffed Tesla on a $50,000 bonus that he had promised, Tesla walked out without a word, eventually finding his way to Westinghouse's door.

AC/DC: It's More Than Rock 'n Roll

Throughout the 1880s, Thomas Edison and George Westinghouse waged a private war to determine whether the electric lights in New York City (and eventually everywhere else) would be powered with direct current (DC) or alternating current (AC). Edison argued that the inherent danger of electrocution made AC too deadly for general use, while Westinghouse countered that DC could only be transmitted a short distance without resorting to the use of prohibitively heavy wires. The issue was finally resolved in 1891 when Edison's firm merged with AC arc-light producer, Thomas-Houston. The new company came to be known as AC appliance giant General Electric. Edison, who had no control over the new company, never admitted defeat. He simply claimed he had more important things to do than waste his time on electric lighting.

Today, we're still using alternating current for most of our electrical needs, and primarily for the reasons George Westinghouse espoused over a century ago. But what is AC, exactly, and how does it differ from DC?

DC is simplicity, itself. It is merely a steady stream of electrons traveling along a conductor from one place to another. It is the current that powers all the systems in an automobile, and the current that flows from a photovoltaic module (solar panel) to a battery or inverter. It is the only form of electricity that can be stored in a battery.

AC is a bit more complicated. Graphically, it can be pictured as a graceful wave (called a sine, or sinusoidal, wave) that undulates from a positive peak to a negative trough and back again. Essentially, it is an electrical current that reverses
Sine Wave

its direction of flow several times per second (60 cycles per second, or hertz, in the USA, 50 hertz in most of Europe).

If you live in a house that is powered (all, or in part) by a solar array, you will be employing both DC and AC, and a large chunk of the money you'll spend will be for the inverter, a mysterious component that converts the DC from the solar array into the AC needed by your house. If your system includes batteries, the inverter will be located between the batteries and the AC sub-panel used to power critical systems. Without batteries, your solar array will connect directly to the inverter.

Either way, both types of current are necessary for your solar-powered home. So, in addition to doing your part to clean up the environment and save the world, you're vindicating Thomas Edison in a way the clever old inventor could never have imagined.

But why was Westinghouse able to send his AC farther than Edison could send his DC? It all relates to a principle called inductance, and the work done by Croatian immigrant and Westinghouse employee, Nikola Tesla. What is inductance? It's the spooky property of an electrical current to create an electromagnetic field around a wire whenever current flows through it. When DC current flows, the electromagnetic field changes only when the flow is turned on or off. But with AC, the current is continuously changing strength and direction, so it's forever sending pulses of electromagnetic waves coursing back and forth along the wires.

As it turns out, these pulses of electromagnetic radiation can be picked up by nearby wires, and a current can be induced inside them, even if no current is flowing directly through the wire. If, for instance, a coil of wire (let's call it the secondary coil) is wound around another coil (the primary coil), a current will flow in both coils, once electricity is sent through the primary coil. Moreover—and this is the really important part because it is the underlying principle of the transformer—if the secondary coil contains twice as many turns in its windings as the primary, the current induced in the secondary coil will be at twice the voltage. Likewise, if the secondary coil has half as many turns, the voltage will be reduced accordingly.

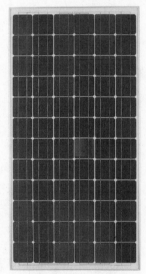

Over its lifetime, this single 175-watt module should produce 12,000 kilowatt-hours of usable electricity. *PHOTO: SHELL SOLAR*

Thanks to the work done by Tesla to perfect the transformer, Westinghouse could begin with a very high voltage at the generating station, and step it down along the way until it became 120-volt house current. Edison, by contrast, had no way to manipulate DC and therefore had to run extremely heavy wires to facilitate the movement of low-voltage DC. Useful things, those transformers, and they owe all their utility to the principle of inductance.

Inductance does have a downside, however, since it's the reason you can electrocute yourself relatively easily with AC, and it's practically impossible with DC—it's those electromagnetic waves that getcha. So, next time you get shocked, blame it on Nikola Tesla and George Westinghouse.

A Look Inside an Inverter

The heart of an inverter is the transformer. It takes low-voltage DC from the batteries or solar array and turns it into the high-voltage AC we use to power our homes. Transformers, however, work on the principle of inductance, a phenomenon that only occurs to any significant degree in the presence of a pulsing (alternating) current. It takes some clever trickery to convince the transformer the direct current driving it is AC.

The magic behind the deception is a configuration called an H-Bridge. Each of the two legs of the H have a transistor switch near each end—four switches in all—and the legs are joined by the transformer in the middle. The two bottom switches control the flow of the negative current from the DC source, while the upper switches control positive flow back to it.

By electronically timing the opening and closing of the switches, the current can be made to flow first one way, then the other, through the transformer. Voilà—alternating current!

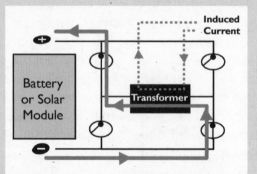

By rapidly opening and closing switches on opposing corners, current flow is reversed and an alternating current is induced in the transformer.

Solar Cells: Teaching Old Electrons New Tricks

The heart of any solar electric system is, appropriately enough, the photovoltaic module, commonly known as the solar panel. A solar panel is usually the squarish unit you buy and mount next to other squarish panels in a solar array, but it can also be in the form of a roof tile, or even a standing-seam roof panel. Whatever form it's in, commercial panels are typically made of single-crystalline or poly-crystalline silicon, non-crystalline (usually called thin-film or amor-phous) silicon, or a hybrid of these. Any way you cut it, slice it or dice it, it comes out silicon.

A solar panel is not the basic unit, however. All solar panels are made up of a number of solar cells, each producing around 0.5 volts of electrical potential. Within the matrix of the panel, these cells are connected in series to achieve the desired voltage, usually in the range of 16 to 60 volts.

Keeping this in mind, we can now answer the one question burning so hot in your mind right now, namely: how do solar cells turn sunlight into electricity? Let me indulge you.

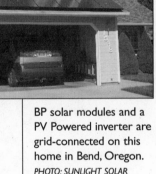

BP solar modules and a PV Powered inverter are grid-connected on this home in Bend, Oregon. *PHOTO: SUNLIGHT SOLAR ENERGY*

What's So Great About Silicon, Anyway?

Largely because of the ubiquity of computers, anyone who has not been living in a cave for the past two decades knows that silicon is a semiconductor of electricity—it sort of allows an electrical current to pass through it, but hardly with the facility of copper or aluminum. Oddly enough, it's this quasi-standoffish attitude of silicon that makes it so useful in the manufacture of solar cells.

Doug's Detour

Solar energy depends on nuclear power! It's true. PV modules are just long-distance antennas for nuclear energy. The power plant, however, is a relatively safe 93 million miles away. It's called the Sun.

Chemically, silicon has 14 positively charged protons, and 14 negatively charged electrons. This would seem to be a happy arrangement, if not for the fact that it has room for four more electrons in its outer energy level. How does it get them? It could snatch four passing electrons from somewhere, but there would be no protons to hold them in place, so the kidnapped electrons would soon escape. So instead, it borrows them from other silicon atoms, forming a crystal lattice in the process (except in the case of amorphous silicon). In this crystal, every atom of silicon is attached to four other atoms of silicon and they all share electrons. In other words, every silicon atom has the four extra electrons it wants, with no net charge, since the protons in the crystal exactly balance out the electrons. It's a really cushy setup.

In fact, it's far too cushy for our purposes. Happy silicon with happy electrons is pretty useless if we want it to do any work. We need to stir things up a bit. How? By adding impurities to it. Say, a few atoms of phosphorus. Phosphorus has five electrons in its outer energy level, so if it is introduced into the silicon crystal lattice (in a process called doping), that fifth electron will be frantically looking for a place to fit in. Now we have an unhappy electron, and that makes us happy.

But we're just getting started. A melancholy electron wandering aimlessly in search of a home doesn't do us much good. We need to give this electron a purpose—something it can aspire to. We do this by making another silicon crystal, this time doping it with a different impurity, such as boron. Having only three electrons in its outer energy level, a boron-doped silicon crystal will have empty spaces where electrons could be, but aren't. These empty spaces are called holes, and each of these holes would like to have an electron to call its own.

Are you beginning to see where this is leading?

Our phosphorus-doped silicon is called n-type, in honor of the extra negative electrons, and the boron-doped silicon is called p-type for the extra positive holes. Now, if we take our silicon crystals, slice them into thin wafers, and carefully join the n-type wafer and the p-type wafer, something interesting happens, and we're getting very close to having a useful electronic device.

Life at the P-N Junction

At first blush you might think that all the extra electrons in the n-type silicon would zoom across the p-n junction (the place where the two opposite types of silicon meet) and fill in all the holes in the p-type silicon, but it just doesn't happen that way. Oh, a lot of them start out fast enough, but quickly begin to have second thoughts. *Sure,* an electron soon realizes, *there may be a nice cozy place for me on the p-side, but my faithful proton is still on the n-side. I'm so confused.* It's a bit like young love.

The important thing to remember is that even though n-type and p-type silicon have extra electrons and holes, neither type, alone, has any net electric charge. In both cases there are just enough electrons to balance out the protons. But once the two types of silicon are joined and the rush across the border occurs, that quickly changes. Every time an n-type electron jumps through the p-n junction and fills in a p-type hole, a negative charge is created on the p-side, while a positive charge springs up on the n-side in the place where the electron was, but no longer is. Once everything settles down, we find that there is a great gathering at the p-n junction, with negative electrons lining up along the p-side, and positive holes lining up along the n-side. This creates an electrical equilibrium, and if we left things like that, the p-n junction would be a really boring place.

But we're not through yet, for now it's time to finish building our solar cell. To do this, we need to crisscross the surfaces of each of our silicon wafers with electrically conducting channels.

Pass Me That Really Good Silicon Glue

Have you been wondering how they attach p-type silicon to n-type silicon? Glue, maybe? The fact is, it's all the same silicon wafer, oppositely doped on either side by a painstakingly exacting process that gives the single piece of silicon the same properties as two perfectly joined pieces.

Manufacturing a PV cell.
PHOTO: SHELL SOLAR

Handy Formulas

Volts x Amps = Watts

Watts ÷ Amps = Volts

Watts ÷ Volts = Amps

Definitions

One watt delivered
for one hour =
one watt-hour

1,000 watt-hours =
one kWh (kilowatt-hour)

This will provide an easy path for the electrons to travel along, once we add the magical ingredient, sunlight.

When a photon of light of the right energy and wavelength strikes an electron hanging out with all of his buddies on the p-side of the p-n junction, the electron is instantly imbued with a jolt of energy and is suddenly free to move around. Where does it go? It can't go any farther into the p-side; there's quite a crowd there already. So instead it uses all this free energy to make a beeline back to the n-side. And, with a little luck, it will be picked up by one of the conductors on the surface of the n-layer and sent through an electrical circuit.

Once the process begins, the electrical equilibrium at the p-n junction is hopelessly undone and the proverbial floodgates are opened. In an instant, multitudes of electrons that were just moments before hanging out at the p-n junction are whisked out of their silicon Shangri-la, drawn through the windings of a washing machine motor or the filament of a light bulb, and unceremoniously dumped back on the p-side of the solar cell, totally exhausted. But, like battered and beaten heroes in a video game, all they really need is a little nourishment—a single photon—to be right back in the thick of things.

A pole barn in upstate New York is used as a grid-tied 10kW power station for a farmhouse located about 150 feet away. The 64 Kyocera solar modules serve as a functional roof for the storage barn, and four SMA Sunny Boy 2500 inverters and remote control are located at the farmhouse. *PHOTO: TRIANGLE ELECTRICAL SYSTEMS*

The completed circuit is the key to making the whole thing work. Since a solar cell acts as a diode—only letting the current flow from the p-side to the n-side—it wouldn't produce electricity for very long without a fresh supply of electrons continually re-entering the solar cell from the p-side. That's why all solar panels have positive and negative terminals. The electrons flow out of the negative terminal which conducts electrons from the n-type silicon, through the load (the above-mentioned washing machine or light bulb) and back into the p-type silicon via the positive terminal.

It's Practically Efficient

This, then, is how sunlight is harnessed to brew your coffee and run your big-screen High-Definition TV with quad-surround-sound. It's sunlight, in fact, that's providing the power to run the computer I'm using to write these words. It spares me the anguish of slowly going buggy while pecking away on a manual typewriter.

How efficient is the process? While efficiencies of over 40% have been achieved in the lab, most commercial panels operate at a peak efficiency of around 15%. This means that 15 of every 100 photons send electrons careening through the circuit. The other 85 are either absorbed as heat, or reflected back into the atmosphere.

Is that sufficiently efficient? Sure; from a practical standpoint, it's plenty. Many people—myself included—harvest all the power they need with a 100 to 200 square-foot solar array. That's less than the roof area of a small toolshed.

The only remaining questions for PV panels are what type should you buy, and where should you put them? But before we get to that point, we need to discuss the different types of grid-tied systems, and which one will work best for you. ❖

The 3% Solution

Before you scoff at the seemingly paltry 15% efficiency of most PV modules, you should stop to consider that green plants—botanical marvels that have been in the solar business a whole lot longer than we have—are only 3% efficient at converting sunlight, H_2O and CO_2 into sugar. And yet they thrive, supporting not just themselves, but all of us non-solar life forms too.

Decathlon Marathon

It was time this house finally found a home. As one of 14 entries in the 2002 Solar Decathlon competition, the CU Decathlon house was conceived and constructed by a student group on the University of Colorado campus in Boulder. From there it was disassembled and shipped to the National Mall in Washington D.C. for the contest's 10-part competition, then back to Boulder where it was put up for sale as the competition's overall winner. The buyer was, appropriately enough, Dr. Ronal Larson, former Chair of the American Solar Energy Society (ASES). Ron moved the house to where it now sits, on a panoramic site on Lookout Mountain, west of Golden, Colorado.

In its original design (as mandated by the competition), the house was to be a 650-square-foot off-grid home office, capable of indefinitely maintaining its heating, cooling, lighting and electrical systems without any form of outside energy,

and still have enough energy left over to charge the batteries of an electric car. Ron thinks the house can do exactly that, even though he expanded it to 2,700 square feet before he and his wife, Gretchen, called it home.

The home's 7.2 kW solar array can produce around 30 kWh of electricity on a sunny day, which is far more than needed. The two Outback MX60 charge controllers first top off a bank of 32 L-16 flooded lead-acid batteries, then the dual Xantrex SW5548 inverters send what remains into the electrical grid (a more worthwhile undertaking than in previous years, now that Colorado allows net metering and time-of-day rates). Heat for the hydronic in-floor heating system is gathered year-round with the original array of evacuated solar-collection tubes and four add-on standard hot water panels working together to heat 10,000 gallons of water stored in two insulated concrete tanks along the north wall of the walkout lower level. A pair of secondary-combustion wood stoves serve as backup during winter months.

What remains? A plug-in hybrid car he and Gretchen can employ to exercise the home's beefy battery bank.

The 2002 Solar Decathlon winner (top) is now part of Larson's home (left). *PHOTO: LAVONNE EWING*

CHAPTER FOUR

You Want Batteries With That?

Grid-tie systems don't need batteries to function, but if you want instant emergency back-up power and the ability to keep using any available solar energy when the normal utility power fails, you'll be needing a battery bank. This is the major fork in system design, because much of the hardware required for battery-based systems is different from what's required for direct-intertie systems. Let's give a quick overview of your options.

Direct Grid-Tie Systems (without batteries)

Direct grid-tie (or intertie) is the simplest, least-expensive, and most efficient way to put a solar electric grid-tie system together. It's the path that about 98% of system owners have chosen to date. Direct-intertie systems treat the utility

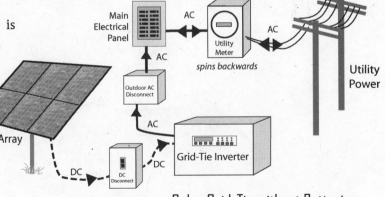

Solar Grid-Tie without Batteries

grid like a battery to absorb any extra, or make up any shortfall of energy. However, if utility power fails, your solar electric system shuts down—even if it's sunny daytime. Direct-intertie systems don't have any storage mechanism, other than the grid, and if the grid fails, these systems will shut off almost instantly for safety. Direct-intertie probably won't work, or at least shouldn't be asked to work, with a back-up generator either. That's too small a "grid," and the generator probably won't react fast enough, or accurately enough, to keep the inverter from shutting off anyway.

Battery-Based Grid-Tie Systems

Battery-based systems use a special inverter that can manage batteries and grid power simultaneously. So long as grid power is available, these inverters act pretty much just like a direct-intertie inverter,

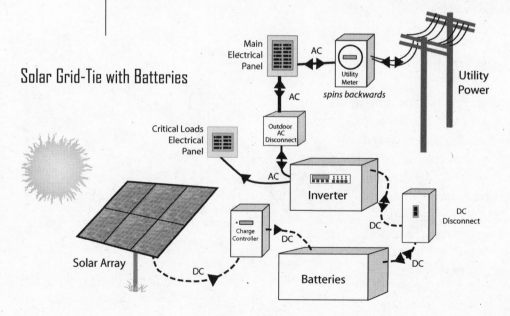

Solar Grid-Tie with Batteries

Main Electrical Panel

AC

Utility Meter
spins backwards

Utility Power

AC

Critical Loads Electrical Panel

Outdoor AC Disconnect

AC

Inverter

DC Disconnect

DC

Charge Controller

DC

Solar Array

DC

Batteries

DC

passing any extra energy off to the grid, or letting the grid make up any shortfall. But if grid power fails, the inverter switches instantly to battery power (well, it actually takes about 20 milliseconds), and continues merrily on its way running anything connected to the critical-loads electrical panel on the inverter's output. These systems tend to be about 2%–4% less efficient than direct-intertie systems at delivering the available PV wattage to your house. That's the price for keeping a set of batteries charged and ready to step in instantly to support selected circuits in your house. The size of your PV array, and/or how much household stuff you want to keep running, determines your inverter size. The size of your battery bank is determined by how long you want to run that household stuff when the grid fails. Battery banks can be very small, starting out at $200–$300, or very large, exceeding $20,000, depending on your needs and budget.

You Can't Take It All With You

Battery-based systems usually don't try to back-up your entire household. We pick a few important circuits, like the fridge, furnace, office, a few lights and live with it. This keeps the battery bank more reasonably sized, and lets the system run longer on what power is available. These systems could certainly be sized to run your whole house, and a system like that would sure make your installer's month a happy one. Americans don't tend to be energy conservative, so for most folks that would be a very expensive convenience. Back-up generators can easily be incorporated into battery-based systems for longer-term use. To make up any energy shortfall, the generator will probably only need to be run for a few hours every few days to recharge the batteries. There's no reason a system like this couldn't continue for months or even years without utility power.

The back-up abilities of these inverters is nearly instantaneous

The Right Tool for the Job

Batteries are built with different chemistries depending on what type of service they'll see. Automotive batteries enjoy a short, ugly life if put into off-grid deep-cycle service, for instance. If regular wet-cell deep-cycle batteries are put into emergency back-up service, a similar mismatch happens. It takes a special kind of battery to thrive under emergency back-up conditions. And that battery tends to be a sealed type.

with most types of grid failure. They take over so fast that often you won't even be aware that the grid has failed. This is good and bad. Good, because your computer won't crash. Bad, because if you don't know you're running on the back-up batteries, there's no reason to be conservative with power use. Switching time is usually under 1/100th of a second. For most home computers this is no sweat. The computer power supply can usually coast through at least 1/20th of a second. However! Be forewarned. No back-up inverter manufacturer

Why Your Back-up Inverter Is Not a UPS

A UPS, or uninterruptible power supply, is the $100 gizmo you plug your home or office computer into (if you or your boss is smart) that keeps your computer from instantly crashing when the grid power hiccups. If you opt for an intertie system with batteries, you'll have a system that seems to just be a great big UPS. After all, it takes over instantly if the grid goes down, right? Well, yes…most of the time. However, there are some kinds of grid failures that will cause your backup inverter to shut down. Let me tell you a tale of woe.

Back in the mid '90s when the first generation of Trace Engineering SW-series inverters came out, Real Goods set one up as a UPS for their phones and computer systems. If the utility failed, we'd have two to three hours of backup battery power, and the ability to plug in a generator to keep going. We figured we had a big powerful UPS for cheap.

Months later, across town, while installing major power for a large shopping center, the contractor had one of those "oops!" moments that very briefly resulted in many of the town's

120-volt circuits becoming 208 volts. Lights flared, but major breakers opened quickly, sparing most folks any damage. At Real Goods, our lights flared, and then everything went back to normal…uh, except that the phones and computers weren't working! For a mail order company in the 1990s, that's the equivalent of "dead."

The SW inverter, upon seeing over 170 volts input, had simply shut off to protect itself. Nice for the inverter, not at all nice for the Unix-based main-frame computer it was supposed to be protecting. Very bad things happen to Unix computers if they're crashed unexpectedly. It was hours before the Information Services crew got our ordering system up, and weeks later, plus hundreds of hours of overtime, before it was right again.

The moral of this story is, if you don't want to crash your computer, buy yourself a $100 UPS, even if you've already got a battery-based backup system.

— *Doug Pratt*

claims that their units will act as an uninterruptible power supply (UPS). There are some kinds of particularly nasty power failures that can cause the inverter to shut down to protect itself. I know. I've been there *(see story on the previous page)*.

It Costs Extra and Doesn't Last Forever

Batteries are the one component of your system that will wear out and need replacement, typically at about 10-year intervals, but it depends on the quality of your battery bank. Batteries will last from 5 to 20 years, with the difference mostly in initial quality and size. Batteries are one of the few things in life where bigger really is better. Larger banks with 15- to 20-year life expectancies are available...for a price. Batteries are also going to need two to four hours of attention and maintenance on a yearly basis. On average you'll find that with the different inverter, extra controls, safety equipment, plus the cost of the batteries themselves, a battery-based system will probably cost about $5,000 – $7,000 more than the same PV array configured for direct grid-tie.

A PV system using Uni-Solar's peel-and-stick thin-film modules on metal roofing. The dark areas are the solar modules. *PHOTO: DC POWER SYSTEMS*

Which Is Right For Me?

Batteries or not? The answer depends mostly on how reliable the grid is in your area, and how much you'll suffer without it. If you can practically depend on regular, lengthy power outages (the hurricane-prone coastal regions of the southeastern U.S. come to mind), then back-up batteries are for you. If you only foresee one or two outages a year, and they're usually brief, then there's little incentive for the extra cost, maintenance, and eventual replacement of a battery bank. Buy a good UPS for your computer, and maybe a propane-fueled back-up generator if you're really concerned. Put your money into more PV wattage instead. It'll do you more good.

How A Backup Power System Saved My Bacon (and my Turkey)

It was 2:00 p.m. on Thanksgiving Day and all twelve members of my local family were at my house when the power went out. Did you know that modern pilotless gas ovens don't work without electricity? It's a fact that about two-thirds of our small town of 15,000 souls was about to learn, because the power stayed off for most folks until about 9:00 that evening. Suffice it to say, there were many ruined dinners and short tempers that day.

At my house? Thanks to some excess paranoia concerning Y2K, I had equipped myself with an inverter that would plug into the bat-tery bank of my electric vehicle. I had 10 kWh of stored energy sitting in the driveway, and the means to deliver it to my house! No problem!

Within five minutes I had the inverter plugged in and the house switched over to backup mode. The silent inverter happily ran the oven, the fridge, the furnace, the microwave, some lights, even the TV for traditional football. We had the only lights on in the neighborhood, and no generator noise. We weren't the least bit inconvenienced and a memorable dinner arrived on time.

— Doug Pratt

Seal Them Up!

We strongly recommend sealed batteries for emergency back-up battery banks. Why? A couple reasons. 1) We've found from hard experience that batteries tend to be forgotten and neglected, and emergency battery banks especially so. Wet cell batteries need watering every three to four months and, despite all the best intentions, they will be forgotten. This leads to sharply reduced life expectancy and performance. 2) Wet cell batteries are like the muscles of your body. They need exercise in order to stay healthy. If wet-cell batteries aren't cycled regularly they become stiff and chemically resistant. They won't be able to deliver the power when they're finally asked to. Imagine what kind of shape you'd be in if you spent two years strapped to a bed and then got thrown out the door to run a marathon. Yeah, ugly isn't it?

Sealed batteries never need watering, and their chemistry and construction is tweaked a bit because they expect to spend long periods doing nothing, and then suddenly get asked to work their heart out. They still won't be real happy about it, but they'll tolerate it fairly well. Sealed batteries are more expensive than conventional wet cells, but for emergency back-up, this is money well spent! *(Read more in chapters 5 and 11.)* ❖

Sun-Xtender® sealed battery by Concorde.
PHOTO: CONCORDE BATTERIES

A 5.76 kW array was installed on this horse barn in Woodstock, Vermont in 1999. Note the steep PV mounting angle for northern latitudes. The grid-tied system has 48 BP solar panels, dual Xantrex SW5548 inverters and 24 Surrette batteries for 35 kWh of storage.
PHOTO: GROSOLAR

Doug's Disappearing Solar PV System

The 48 Uni-Solar peel-and-stick panels on my shop are nearly invisible, but the 20 kWh per day of electricity (on average) that it provides is quite noticeable, as is evidenced by the zero electric bills from my utility company.

We decided to install a battery-based system because lengthy power outages aren't uncommon in our rural area, and with the nice California rebate we could afford to spend a bit more for the security. In my power room foreground you see the sealed batteries with a wooden cover so nothing can get dropped on them. On the wall, there's pair of Outback 3,600-watt inverters with a DC box and a pair of the superb MX60 charge controllers on the near side, and an AC box on the far side for all the wiring and safety gear. Further down the wall there's a pair of standard circuit breaker boxes. One is for utility power, the other for inverter power. Note they're connected by a gutter box so individual circuits could be easily moved if we change our mind about what wants backup power.

The energy-efficient, passive-solar, all electric home that goes with this large shop uses a geothermal heat pump for heating and cooling. It's the least expensive way to heat or cool a house, even when paying regular electric rates. I'm using free solar power to run it. My electric bill is about $100 per year, and I'm having zero impact on resources available to my children and others. I can make electricity; I can't make propane or natural gas. This all makes my eventual retirement look a lot more secure.

— *Doug Pratt*

PHOTOS: DOUG PRATT

CHAPTER FIVE

Sizing Your Solar / Wind Electric System

Before I began writing a monthly column about people who live with solar and wind energy, I figured everyone should install the biggest renewable energy (RE) system they could afford, right up to the point where all of their electrical energy needs were covered most of the time, without resorting to a backup generator or the power grid. Though I still think it's a good idea, I've come to the realization that most people think differently than I do about this matter. I've met people with huge RE systems who, because they refuse to adopt a more energy-efficient lifestyle, still rely heavily on grid power, or, if living off-grid, generator power. Others are happy with one or two panels and a couple of batteries to run a small TV or computer, and simply do without electricity for everything else. To each their own.

Since most of us install our systems with a particular goal in mind—to provide backup power during grid outages, for instance—sizing such systems requires a fair bit of planning based on reliable data.

Different Strokes for Different Folks (with different systems)

Direct grid-tie systems without batteries do not provide any power when the grid goes down, so they can be as big or as small as you

Hidden Savings

If you generate one kWh of your own electricity, you save much more than one kWh from the utility company. Over 3.3 kWh of energy were consumed to deliver that one kWh to your home.

Carbon Dioxide Carriers

Replace five 60-watt incandescent light bulbs (each burning 5 hours per day) with 12-watt compact fluorescent bulbs, and you'll offset the burning of nearly 440 pounds of coal and over 860 pounds of CO_2 going into the atmosphere **each year**. If a million people did this for 20 years, the CO_2 not sent up the smokestack would outweigh 77 Nimitz Class aircraft carriers, each boasting a displacement of 102,000 metric tons!

desire, or as your pocketbook permits. Why are you installing this type of system? If it's because you want to pitch in and do your part to offset the burning of tons of coal, or to reduce the size of the growing mountain of radioactive wastes produced by nuclear fission, then your decision is a simple monetary one: buy whatever you can afford. On the other hand, if you hope to see a credit on your bill from the power company now and then, you'll need to do some figuring.

Are you planning a system with batteries? If so, the planning stage becomes more critical, since you will need to know, at a bare minimum: 1) How much power is consumed by the appliances and home systems you wish to keep operating when the grid goes down; 2) How many watts of solar capacity it will take in your particular geographical area to keep the batteries at an acceptable state of charge during a power outage; 3) How big of an inverter you will need to provide power to the loads it will be running; and 4) How big of a battery bank will be required to power your home during a power outage.

You'll also need to know where you're going to put all the components, since batteries take up a fair bit of room and need to be as close to the inverter as possible.

But I'm getting ahead of myself.

Watt(s) You Use versus Watt(s) You Need

A quick glance at your monthly bill from Planet Power Conglomerated will tell you how many kilowatt hours of electricity you used in any given billing cycle, and a little math will give your daily usage. That's the easy part, and if you're willing to spend $7–$9 per watt on the solar panels, installation and hardware required to maintain your current degree of consumption, then simply plug that figure into the system-sizing table in the appendix and don't give it another thought.

But before you get the shock to your wallet that option delivers, you might want to examine how you use your electricity, since your house is probably full of energy-wasting devices and appliances. This is because electricity has traditionally been so cheap that manufacturers have had little incentive to incur the added expense of engineering and producing energy-efficient products.

The biggies are obvious: refrigerators and freezers, air conditioning, forced-air heat, electric ranges, water heaters and clothes dryers. But there are also many little things we don't usually consider that, taken as a whole, can make a big difference. Light bulbs, for instance. Every 60-watt incandescent bulb you replace with an equivalent compact fluorescent bulb using only 12 watts will save 0.048 kWh for each hour it's turned on. If you switch out five light bulbs that burn an average of 5 hours per day, you can knock 36 kWh per month (1.20 kWh/day) off your electrical consumption. This is enough energy savings to run a reasonably efficient (note that I did not say "expensive super-efficient"), off-the-shelf refrigerator 24/7 and still have a few kilowatts hours to spare at the end of each month.

How many electrical devices and appliances do you have around the house that come with heavy, black plug adapters? Those adapters or power cubes are actually rectifiers used to convert high-voltage AC to low-voltage DC. You will find them on laptop computers and many peripherals, cell-phone chargers and other chargers for portable devices, TV cable boxes, lamps, etc. They are also used internally in stereos, computers, battery chargers for portable tools, and so on. They all have one property in common: they draw power continuously—usually in the one- to four-watt range—for as long as they are plugged in, even if the power to the appliance is turned off. In fact, power companies don't call them power cubes; they officially call them "power vampires." How can you tell if your pet vampire is using electricity? If you don't have a usage meter to plug it in to—which is really the best way—simply touch it. If it's warm, it's using electricity. How do you

Energy Pigs

Not-so-old fridges are energy pigs. A new Energy Star model uses half the energy of a fridge built in 1993.

Demand Keeps Growing

World energy consumption is projected to increase by 59% from 1999 to 2020. Much of that growth is expected in developing countries.

SOURCE: SOLARBUZZ.COM

stop these things from using electricity? Just plug them into a power strip and turn off that power! The savings over the long haul can be substantial.

Electric clocks are also accomplished watt thieves, usually drawing one to three watts continuously. I know it doesn't sound like much, but over the course of a year, a 3-watt clock will consume in excess of 26 kWh of electricity. That's over a day's electrical consumption for an average house, just to power a timepiece that could be traded out for a battery-operated model that will run for two years on a single AA battery! I've often wondered what the designers of battery-operated clocks know that the plug-in guys don't.

Okay—so much for the little stuff. Now you have to ask yourself how many of the biggies you're willing to part with to make your excursion into solar energy a worthwhile venture. Below are the average power-consumption values for some downright piggish electrical appliances (*see the appendix for a longer list*):

Electric clothes dryer	4,000 - 6,000 watts
Electric range	1,200 - 2,600 watts (per burner)
Electric water heater	4,500 - 5,500 watts
Central air conditioning	2,000 - 5,000 watts

Does it take an hour to dry your clothes? Kiss goodbye to 4 kWh of electricity. This is roughly equal to the average daily output of a 1,200-watt solar array in a sunny climate at mid latitudes. Fortunately, a natural-gas or propane-powered dryer uses 1/10th that amount of electricity, and burning natural gas or propane in your home is far more environmentally benign than burning coal to make electricity at some distant power plant. The stove and water heater can likewise be replaced with gas models. Was your fridge manufactured before

1993? If so, it uses twice the energy of a new Energy Star model. To see just how efficient the new models are (in all configurations and sizes), check out *www.energystar.gov*.

This leaves us with the biggest pig on the whole farm, central air conditioning. How badly do you need it, really? Could you go through a summer with a new evaporative cooler, or a few fans, instead? Before you answer, you might want to complete the system-sizing worksheet in the appendix, with and without the central air. The difference could give you the chills.

Building a new home? Look into ground-source or geothermal heat pumps *(see chapter 15)*. They are by far the least expensive cooling and heating sources now available. Of course, by designing your home to take advantage of passive solar heating and lighting, then building it tight and insulating it well, you'll set yourself up for a lifetime of energy savings *(see chapter 13)*.

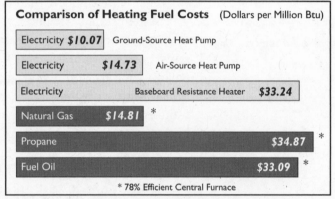

Comparison of Heating Fuel Costs (Dollars per Million Btu)

Electricity **$10.07**	Ground-Source Heat Pump	
Electricity	**$14.73**	Air-Source Heat Pump
Electricity	Baseboard Resistance Heater	**$33.24**
Natural Gas	**$14.81**	*
Propane		**$34.87** *
Fuel Oil	**$33.09**	*

* 78% Efficient Central Furnace

Fuel Costs are late 2008 averages: Electricity $0.113 / kWh; Natural Gas $1.55 / therm; Fuel Oil $3.58 / gallon; Propane $2.48 / gallon
SOURCE: DEPARTMENT OF ENERGY'S HEATING FUEL COMPARISON CALCULATOR

Sizing the Components: Inverters and Charge Controllers

Now that you know where most of your energy is going and what you can do to slow the flood of watts to a manageable stream, it's time to think about system size. For direct grid-tied systems without charge controllers and batteries, your only real concern is making certain that your inverter is big enough to handle the wattage generated by your solar array. Are you thinking of adding more solar panels down the road? You could get a bigger inverter now, or you could buy an

www.energystar.gov

is an excellent resource that lists energy consumption of appliances and much more.

"Nobody has ever lost money on energy efficiency. There's so much low-hanging fruit available that it's mushing up around our ankles."

AMORY LOVINS, CHIEF SCIENTIST, ROCKY MOUNTAIN INSTITUTE

inverter well-suited to your initial array and add a second inverter when you get around to increasing your solar capacity.

For systems with batteries, your initial planning is more critical. Once you determine the size of your solar array *(see the worksheets in the appendix)*, you can match it up with a proportionately sized battery bank and charge controller. The inverter, on the other hand, is sized to meet the demands on the AC side of things, so unless you add onto your house or increase the loads you'll be running, the inverter you begin with will more than likely serve you well, even if the rest of your system suffers from occasional growing pains.

Charge controllers—at least the ordinary garden variety—are fairly inexpensive and if you need to add another one when you enlarge your array, you won't be out too much money. But if your initial array is capable of, say, 35 amps, and you suspect you may want to add onto it someday, your money might be better spent in buying a 60-amp charge controller rather than a 40-amp model.

Do You Fade in the Heat?

PV modules fade too. Crystalline modules are test-rated at 77°F (25°C) and their power output diminishes by around 2.7% for every 10°F rise in temperature (5.0% per each 10°C). Additionally, since modules are dark-colored, they operate at higher temps—at least 20 degrees—than the ambient air temperature. If it's 80 degrees and sunny outside, your modules will be running at well over 100 degrees. So on hot summer days you can figure their power output will fade at least 10% from specs.

The flip side of this behavior is that PV modules **increase** their output by the same proportions as it gets colder. On a clear winter day when your modules are operating near freezing, you'll enjoy a 10% output gain.

For amorphous or thin-film PV modules the change per degree is about half as much. If you live in a hot climate or your power needs peak in the hottest times of the year, then thin-film modules are a smarter investment. But if yours is a cooler climate or your power needs increase as the weather grows colder (the typical off-grid system), then crystalline modules are a better bet.

How Many Batteries?

A battery bank is a bit tougher to size. Ideally, you will want to have enough battery power to get you through a world-class power outage in the midst of a cloudy spell, without running your batteries below 50% capacity. But this isn't always realistic. Batteries are the one part of your system that you KNOW is going to wear out eventually. It doesn't make good financial sense to invest too deeply here. Even the best batteries will fade away after 15 to 20 years, while your PV array will still be going strong at 40 years.

Batteries take up a lot more space than other indoor components. For instance, MK's 8G31 12-volt sealed gel-type batteries—a tried-and-true favorite among grid-tied renewable energy enthusiasts—take up roughly one square foot of floor space for every 2 kWh of usable capacity. So, if the systems you're running require 10 kWh per day and you want enough battery capacity for 3 days, you'll need to find at least 15 square feet of unused space to put your batteries.

And if you should want to add more batteries later? Your batteries will need to be purchased in sets of two or four and the size of the sets will be based on your system voltage. Why? Battery-based systems usually operate at 12, 24, or 48 volts, the latter two being by far the most common. If your batteries are of the 12-volt persuasion (as with the above-mentioned MK batteries), you will need to wire them in series to achieve the system voltage you want: a 24-volt system requires that batteries be installed in sets of two, while a 48-volt system takes batteries in sets of four. If you instead opt for 6-volt batteries, these numbers will be doubled. The only way around this mathematical conundrum is to buy an MPPT charge controller that allows you to run a 24-volt system from a 48- or 60-volt array (*see the section on MPPT charge controllers, pages 122–123*).

Doug's sealed 48-volt / 555-amp hour battery bank can deliver a very comfortable 13.5 kWh at 50% depth of discharge, and requires 5 minutes of maintenance per year. It's shipped in 6-packs with solid soldered interconnects between cells (visible in the center), which you see here with nice, safe plastic covers over them. Non-removable caps provide a one-way vent if the batteries are over-charged. Life expectancy is about 15 years. *PHOTO: DOUG PRATT*

System Voltage

In the early days of solar, most systems were on backwoods shanties and comprised of a couple of PV panels and two or four batteries. These systems usually operated at 12 volts because that's the voltage needed by the RV lights most people were using back then. But things have changed. Today's components are far more sophisticated and, watt for watt, are considerably cheaper. We want to run more than just a few lights and a small TV; we'd like to rival the wattage we buy from the power company.

For reasons discussed earlier *(see chapter 3)*, higher voltage translates to higher efficiencies. The wires running from a 48-volt array need only be 1/4th the size of the wires needed for a 24-volt array, and 1/16th the size required by a 12-volt array. Likewise, a 48-volt inverter will operate more efficiently than a 24- or 12-volt inverter. And these days most large solar panels are wired for 24-volt operation, which means it only takes two panels wired in series to make a 48-volt array.

The higher the voltage the better, then. Most direct grid-tied systems operate at 300 to 500 volts, depending on which inverter you choose. For battery-based systems 24 and 48 volts is the norm, with 48 volts as the standard.

Are you planning a battery-based system? Do you have the room (and the funds) to increase the size of your battery bank four 12-volt batteries (or eight 6-volt batteries) at a time? Then go with 48 volts. Voltage is good. ❖

Sharp makes angled solar modules to fit hip roofs, as shown here.
PHOTO: SHARP

CHAPTER SIX

Where Does Everything Go?

Where can you—and can't you—put your solar array, batteries, inverter(s) and charge controller(s)? I'd like to say simply that everything except the solar array has to go inside, but that's not exactly true, so here it is, caveats and all.

Where Should I Put My Solar Array?

Since your solar array will be located outside, space is usually not a problem unless you've got a huge array and a tiny house. Crystalline silicon panels will take up about one square foot of space for every 10–12 watts of power they produce; amorphous silicon panels over half-again that much area. If you've calculated you'll need a 3,000-watt array, you can mount it on 300 square feet (28 m²) of roof if you are using crystalline panels, or 450–550 square feet (42–51 m²) of roof for amorphous panels, or standing-seam solar roof panels. Since most residential roofs are over 1,500 square feet (140 m²), finding a place for your array should not be a problem if a good portion of the roof faces south. Likewise with ground- or deck-mounted arrays.

There seems to be an almost universal misconception among those who have never had PV equipment that the array has to go on

Face the Sun

In the northern hemisphere, your PV array wants to face due south, but there's some wiggle room. Within 25 degrees of true south, you'll still collect 98% of the sun's energy.

Standing Up to Hail

All UL-listed PV modules (the kind you want to buy) have to pass UL-1703, which states they must withstand a direct 90° strike by a 1½ inch hailstone traveling at terminal velocity (about 52 mph). If you're expecting larger hail, then Uni-Solar laminates are a great choice. Anything that doesn't go through the roof won't hurt them.

the roof. It doesn't. In fact, in many instances you're better off mounting it on terra firma. It's easier to wash the panels when they're at ground level, easier to sweep away snow in winter, and easier to make seasonal adjustments to the tilt angle.

Mounting the array on the outside of a south-facing deck is another option, albeit not a very aesthetically pleasing one. The payoff is in the fact that it's even easier to clean a deck-mounted array than one at ground level. And it will be less likely that the kid you pay to keep the weeds at bay with a weed wacker will send a rock careening at Mach 2 into one of your expensive panels.

But we don't all have big, sunny yards or decks, and for you it may be the roof or nothing. If your roof is composed of composition shingles or standing-seam metal panels, it should be a fairly straight-forward job. It is not difficult to seal-up around the places where the bolts penetrate through a composition shingle roof, and for standing-seam metal roofs there are non-penetrating mounts available.

Wood shingles are a bit more challenging, owing to the fact that they have a tendency to split. For these types of roofs we strongly recommend stand-off type supports with an impermeable flashing that's designed for wood shingles. You might want to consult a good roofer to make sure it's done right.

Tile, either clay or concrete, is another step up the ladder of

Easy snow removal for deck-mounted and ground-mounted solar arrays. *PHOTOS: LAVONNE EWING*

Pole-mounted arrays adjust easily to the optimum sun angle. *PHOTO: LAVONNE EWING*

Avoid the Shade

Shading is far more important than direction. If turning east or west helps you avoid shading, go for it!

Tilt to Meet the Sun

The sun passes through 47 degrees of latitude in its journey from the Tropic of Capricorn on the first day of winter (December 21), to the Tropic of Cancer on the first day of Summer (June 21). This means that the angle at which sunlight hits your array is constantly changing. And, unless, you adjust your array to keep it perpendicular—or nearly so—to the incoming light, the array's efficiency will wax and wane right along with the seasons.

If you have a ground-mounted array supported from a central pole, you will be able to adjust the angle fairly easily; just place an angle finder on the array, remove a couple of bolts, tilt the array to the desired angle, and replace the bolts into different holes. But how do you know what angle is optimal?

On March 20 and September 22 (the vernal and autumnal equinoxes) when night and day are the same length everywhere in the world, your array angle should be equal to your latitude. Then as the seasons change, you can increase or decrease the angle to ensure your array is always delivering as much wattage as possible.

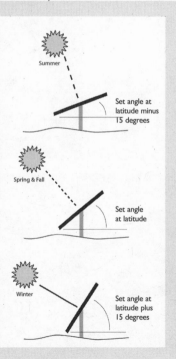

difficulty. As an erstwhile roofer, I can attest to the recalcitrant nature of tile roofs. That being said, they're not impossible; they simply require an extra measure of thought and caution. There are mounting systems made especially for tile roofs *(see chapter 11)*, and in many areas these mounting systems are required.

If you'd rather go about it the old-fashioned way (and local codes permit it), you can have a roofer remove the tile where the supports meet the roof and install a rubberized flashing (or some other membrane-type roofing material). The flashing is attached to the roof deck underneath the tiles that are above the array support, and then glued to the top of the tiles below the support. A fair bit of work, yes, but not terribly difficult for a roofer who's been around awhile. Find one sprouting a few gray hairs—it's a cinch he's not new to the trade.

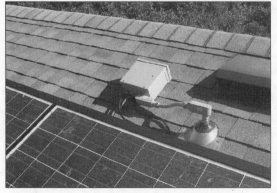

Your PV array is going to last 50 to 80 years, so lay a good foundation. *TOP LEFT:* A securely installed standoff, bolted to a roof framing member. *ABOVE:* All-metal flashings that are caulked around the top. Also shown: a splice bar on the top rail, and a ground wire (on the lower standoff) waiting for PV modules. *LEFT:* Flashing with a no-caulk rubber insert. The junction box will provide space to convert the MC connector cables to conventional (less expensive) wiring before dropping down to the inverter area.
PHOTOS: AFFINITY ENERGY

But I'm Not Allowed to Make Holes In My Roof!

Got a condo or apartment with a big flat roof and a management that says no permanent mounting? You need ballast mounts: low-profile, non-penetrating mounts that are designed to stay in place with just some weights—standard concrete paver blocks work well. Obviously these

don't work with sloped roofs, but they are widely used on commercial and industrial buildings with flat or near-flat roofs. Cost is slightly higher than penetrating mounts, but it makes solar arrays acceptable to building owners, and if you move, you can take it with you.
PHOTO: UNIRAC, INC.

Making Space for the Batteries, Inverters and Other Stuff

How much space does it all take? As we mentioned earlier, MK batteries require roughly one square foot for every two usable kWh of stored energy. Less powerful batteries may take up twice that much space. Bigger, more efficient batteries are available, but they require a commensurately greater investment. Many people find room for batteries under work benches, others build very sturdy shelves to stack them. There are only three requirements for locating your batteries: 1) They need to be as close to the inverter as possible; 2) The batteries cannot, however, be located directly under the inverter; 3) Wet lead-acid batteries must be in a sealed room or box and must be vented to the outside, since they give off highly flammable hydrogen gas

Running Your Solar Circuits Circuitously

Your wire runs might be longer than you anticipated if your meter is located on a distant pole and the local utility requires a solar DC disconnect within 10 feet of it. Rather than doing a marathon wire run from the array, to the meter, to the inverter, many choose instead to install a whole-house disconnect at the meter. *(See chapter 11 for details.)*

when they get worked up (sealed batteries, however, need not be sequestered).

Compared to the batteries and the solar array, the rest is small change, topographically speaking. Even if you're planning a pair of inverters and charge controllers, you should be able to fit it all on a half-sheet (4-foot x 4-foot) of ¾-inch plywood mounted to the wall—

A pair of PV Powered grid-tie inverters (with their covers removed for wiring) mounted on an exterior wall.
PHOTO: PV POWERED

which is far preferable to direct wall mounting of the components, owing to the fact that inverters for battery-based systems are exceedingly weighty beasts—upwards of 100 pounds (220 kg). Besides, wall studs are like cops: you can never find one when you need one.

There is a big difference between direct grid-tie inverters and battery-based inverters when it comes to finding a place to put them. Direct grid-tie inverters are happiest if they're mounted outside, but you've got to find a shady spot for them, since they will limit their output to control their internal temperature, should they get too hot. Unshaded south- and west-facing walls, then, are a bad idea, while north-facing walls are perfect—as long as you don't cover them up just because they're not pretty to look at. They've got to have air. Can't find a suitable outside location for your direct grid-tie inverter? Buy a model with an internal fan; these units can safely be mounted in a garage or utility room.

Battery-based inverters are a different matter. They *must* be mounted in a building or enclosure out of the elements. Does this building need to be climate controlled? No, it doesn't; inverters can handle reasonable extremes in temperature. But you don't want to put one in a barn with dust and snow blowing through the cracks in the walls, either.

Many people find it convenient to locate their components in tightly built sheds beyond the house, then run the AC to the house via a buried cable. It gets everything out of the way, and ensures that they don't have to listen to the hum of the inverter, which can get a bit distracting when it's really working.

The only problem we have with this arrangement is the fact that the batteries have to be in the same general location as the inverter, and battery performance drops with the mercury. How much? Between 80°F (26.7°C) and freezing (32°F or 0°C) short-term power availability drops off by a third. Go down to 0°F (-17.8°C) and your battery efficiency drops to 40%. It's something to keep in mind.

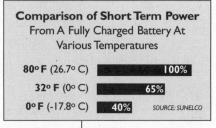

Comparison of Short Term Power
From A Fully Charged Battery At
Various Temperatures

80°F (26.7°C)	100%
32°F (0°C)	65%
0°F (-17.8°C)	40%

SOURCE: SUNELCO

Ideally, then, you'll want to mount your components inside your house for battery-based systems and outside for direct-tie systems. But where inside? Non-living areas in basements are ideal, as are workshops and garages. Basically any dry, secluded area where the hum of the inverter will not be a distraction. Of course, if you locate the inverter in the closet of the guest bedroom, you can practically be guaranteed that your in-laws' visits will be short ones.

A 1.5 kW array feeds a sub-panel on this home office in South Strafford, Vermont. The electricity is then routed to the main house for their non-battery, grid-tied system. *PHOTO: GROSOLAR*

How Far Away Can It Be?

If you decide to locate all of your components in a shed a good distance from the house or, conversely, to place your solar array very far to the south of the house (and the inverter), your only real constraint is voltage loss though the wires. The farther electricity has to travel through a wire, the greater the losses due to resistance within the wire. How much loss can you afford? You absolutely, positively, do not want your voltage to drop by more than 2% from point A to point B. The reason

is that a drop in voltage is accompanied by a corresponding drop in amperage, which will be even greater than the loss in voltage. You're investing a lot of money in your system—there's no point in sacrificing your hard-earned watts to skinny wire.

So, again, voltage is good. A 48-volt array, for instance, will safely deliver 1,200 watts a distance of 184 feet through 1/0 wire; a 24-volt array can only push that much current 46 feet with minimal loss. Of course, if you are planning a direct grid-tied system operating at several hundred volts, then you have a great deal more latitude in where you place your array.

The same is true for distant inverters. What's true for DC holds for AC, so if you run 120-volt AC to the house from an inverter in a shed, you will get better than a four-fold boost in the wire's carrying capacity over the 48 volts from your array. But before you get too heady, remember that you will sending much more wattage from the inverter to the house than from the array to the inverter, since you will have to size your wire to match the surge output of the inverter, a considerable amount of current for a reasonably sized inverter.

The bottom line? Plan ahead. Before you decide to locate your array or inverter on that sunny spot south of the distant knoll, stop to calculate how much you'll be paying for the wire that's going to be delivering the current.

Got Time-Of-Use Metering?

If you're in one of those utility districts that offers time-of-use metering with a significant difference between morning and afternoon rates, then facing your PV array southwest, rather than due south, will probably earn you a much better payback. For instance, in northern California a kWh at noon in the summer is worth 14.6 cents. An hour later that same kWh is worth 29.6 cents. A southwest-facing array will make less kWh per day, but it will knock 30% – 40% **more** off your bill. The utility gets your excess kWh when they really need it, and you get rewarded handsomely for providing it. Everybody (with solar) wins!

Grid-Tie for the Space- or Cash-Challenged

Would you like to go solar but you live in an apartment or condo? Or you can't get past the $10,000 startup investment? Your time has come, thanks to the Enphase inverter. Enphase uses one mini-inverter per PV module, so you can start with a system as small as a single 175- to 225-watt PV module. If you want to expand next year, just buy another module and inverter...or several of them. The only limits are your available space and budget. If you think this sounds really good you'll find a lot of solar pros who agree with you. And the standard 15-year Enphase warranty doesn't hurt either.

There will be some one-time expenses with your first Enphase. You'll need a dedicated 15-amp AC circuit for the array. This will handle your first 12 to 16 modules. If you want more, you can add another 15-amp circuit. The $365 Energy Management Unit (required for system monitoring) will monitor up to 240 modules and talks to the inverters through your AC wiring. No special data wiring required! Just plug the EMU into any convenient outlet. If you have broadband web access, you can plug the EMU data output into your router and send it to Enphase's eye-popping Enlighten Monitoring Service (just a few dollars per inverter per year). You can check your system from anyplace in the world, and if there's a problem it will text or email you.

This 2.8 kW grid-connected carport in Mississippi uses an Enphase micro-inverter with each solar panel. Planning for the future, both 120- and 240-volt plug-ins are located on the center pole. *PHOTO: MISSISSIPPI SOLAR*

Costs will naturally vary with installation difficulty or tidiness required, but at early 2009 prices and assuming the 175-watt PV module, you should budget about $2,000 – $2,500 for the first professionally installed module, and roughly $1,500 for every module thereafter. ❖

Doing What It Takes To Be Self-Sufficient

Yes, living in New York does have its advantages. After Chris and Kimberly Andersen installed a direct grid-tie solar-electric system on their 2,200-square-foot home in Saratoga County, the state promptly sent the installer, Global Resource Options (groSolar), a check for $15,000, a credit which was reflected on the Andersen's bill. Then the state gave the homeowners an income tax credit and also bought down their loan to a comfortable 1.5% interest. At the rate they're cutting their monthly electric bills, the system will have largely paid for itself by the time the loan matures in 10 years.

What did the Andersens and NYSERDA (New York State Energy Research and Development Authority) invest in? An array of two dozen 160-watt BP solar panels on a Unirac roof mount and a pair of Sunny Boy 1800U inverters. During the first year the system has been in operation, the Andersens were able to slice 75 percent off their electric bill, but they are hoping for more. By installing double-glazed low E argon-filled windows, compact fluorescent light bulbs and high-efficiency appliances—the latter also partly subsidized by the state—Chris and Kim hope to conserve enough in the coming years to cut their electric bill to zero.

Why did they do it? Chris says, simply, "We wanted to do what we could be become self-sufficient." Chris spent time in Botswana with the Peace Corps before meeting Kim. The couple—both educators in the public school system—then lived in Paraguay for awhile after tying the knot. These are two places where limitless electricity and modern conveniences are in short supply.

Do they have any desire to emulate the lifestyle of a developing country? Of course not; they just see no reason why they and their two small children should have to consume any more natural resources than necessary.

PHOTO: ANDERSENS

Got Wind?

D o you live out of town on a nice little acreage where you spend more time listening to the wind soughing through the trees than you do listening to your neighbors banging around across the fence? You should consider yourself fortunate; if that ol' north wind is strong enough, it could go a long way toward lowering your electric bills and powering your home during blackouts.

Wind power is not for everyone, of course. In many areas of the U.S., the force of the wind near ground level is too slight to be of practical value for home-based wind systems. And even if you do have usable wind power where you live, several other factors need to be considered before you rush out and buy a wind system.

Yea! The wind is blowing!

Have you spent your life cursing the wind? Installing a wind system is guaranteed to change your attitude.

A Matter of Geography

In addition to the wind itself, you will need to have enough land to safely locate the tower on which the turbine will be placed. Usually, this means an acre or more. Some folks are able to avoid installing a tower by mounting one or more small wind

turbines (500 watts or less) on a barn or other non-living out-building, where the vibration caused by the spinning blades will not shake the plaster off the walls and induce periodic fits of insanity. However, for larger machines (600 watts and up), a tower is a must.

There are two things to keep in mind when searching out a spot to place a tower. First, it should be far enough away from living areas and property lines that no one would be injured (or worse) if the tower fell over, or if all or part of the wind turbine came flying off in a killer wind or microburst—worst-case scenario stuff, in other words. How far is far enough? A good rule of thumb is 15 rotor diameters away from the house. So if your turbine has an 11-foot (3.35 m) pro-peller, you'll want to place your tower at least 165 feet (50.3 m) away.

Secondly, the tower should be high enough to clear any obstacles that might be in the path of the wind. Ideally, the turbine should be mounted at least 30 feet (9.2 m) above the tallest object—tree, building, hill, etc.—within 300 feet (91.5 m). This requirement is to minimize the effects of air turbulence, which is to a wind turbine what a washboard road is to a car. Besides causing extra wear and tear on the turbine, turbulence greatly diminishes the force of the wind.

How Much Wind Is Enough?

So you've got land with an ideal location for a wind turbine, and nei-ther your neighbors nor the local bureaucrats have any objections to a tower. Now you need to determine if you have enough wind at your site to justify the time and expense of installing a wind system. The first thing you will discover is that exact wind data for your particular location is probably non-existent, unless your home is next to an air-port or a military base. But you can still get a pretty good idea what the force of the wind is in your area by referring to the wind maps for your state. The Department of Energy (DOE) maps at the Bergey

Persuasive Math

Ever see a news story about a guy who refuses to leave his home, even after being told his house is about to be deconstructed by a hurricane? I've often wondered what these hard cases would do if someone informed them that a 150-mph wind is not merely 3 times stronger than a 50-mph wind, as they probably imagine, but rather a rafter-rending 27 times stronger.

Windpower website (*www.bergey.com*) assign a wind class number to every square inch of every state. Though these maps are painted with a rather broad brush, they still offer a lot of insight for the wind resources that are available in your area. For further clarification, you should also read the National Renewable Energy Laboratory's Wind Energy Resource Atlas of the United States at *http://rredc.nrel.gov/wind/pubs/atlas/chp1.html*. This is a thorough document that discusses national and regional wind patterns, seasonal variations, and the painstaking methods used to compile the data.

A quick glance at the national map *(see page 170)* will show you that the most paltry wind resources are in the Southeast, while large areas of excellent wind are in the upper Midwest, particularly the Dakotas and the western edge of Minnesota. Good winds can also be found in the higher terrain of both the Northeast and Northwest, and all along the Rocky Mountains.

Another website you might find useful is FirstLook® (*http://firstlook.3tiergroup.com*). It's a free website so long as you don't require any detailed information, but you will have to take a minute or two to set up an account. What do you get for your trouble? Wind speed and solar irradiance data for any place on the globe, so long as you can provide the address or the geographical coordinates. Bear in mind, however, that this information results from extrapolation of existing data from the nearest sites where measurements actually were taken. No one really knows for sure what the wind characteristics are at the top of the tower you haven't erected yet.

To be absolutely sure there's enough wind at your site, you may want to buy an anemometer and monitor its readings for a few months. If you get a fancy recording anemometer, or one with a computer interface, you will actually be able to plot wind patterns over time to determine what the average wind speed is at your location for different months of the year. Just don't forget that your

Small Wind Turbines

The U.S. market for small wind grew 78% in 2008 with an additional 17.3 megawatts of installed capacity.

SOURCE: AMERICAN WIND ENERGY ASSOCIATION

Southwest Windpower Whisper 100. *PHOTO: SOUTHWEST WINDPOWER*

**Tax Credits and
Rebates for Wind?**

Yes, there are financial
incentives for prospective
wind farmers. In many
cases, a home-based
wind system is eligible
for the same tax credits
and rebates as a solar
system. Check the
sources discussed in
chapter 9, as well as your
state energy office listed
in the appendix.

anemometer readings will not be entirely accurate unless you mount the instrument at the same height as your proposed wind turbine. The lower you place it, the less encouraging your results.

Should you go this route, the folks at Iowa State University have compiled a Wind Energy Manual that will tell you, among other things, where best to locate your anemometer and how to extrapolate the data you collect to calculate probable wind speed at different heights above different types of terrain. You can download this handy document at: *www.energy.iastate.edu/renewable/wind/wem-index.htm.*

But you really shouldn't have to sort through mountains of data, or spend a lot of money on wind monitoring equipment because the simple fact is, if you think you have enough wind at your site, in all likelihood you do. My own personal rule of thumb goes as follows:

> *If the wind at your site blows often enough and hard
> enough to annoy you, you probably have enough wind
> to make good use of a wind system.*

However you go about it, there are some surprising facts about wind speed and the amount of power you can hope to harvest from it. For starters, the relationship between the speed of the wind and the power it generates is not a simple linear correlation. What am I talking about? Just this: a 30 mph wind is not, as you might imagine, half-again as powerful as a 20 mph wind—it's nearly 3.4 times stronger! How can this be? It's because the force of the wind increases as the cube of the wind speed: 20 x 20 x 20 = 8,000 while 30 x 30 x 30 = 27,000. If you then multiply either of these products by 0.05472, you will discover the force of the wind in watts per square meter (W/m^2) at sea level for that particular wind speed. This is a tidy arrangement, because it turns out that solar irradiance is also measured in W/m^2, so it's a simple matter to compare the speed of the wind hitting the blades of a turbine with the sunlight that falls on an array.

And how do they compare with one another? Generally, the power of the sunlight hitting the earth (or your solar array) in the middle part of a summer's day at mid-latitudes is equal to a steady wind speed of 26 to 27 mph (42 kph)—about 1,000 W/m^2.

This isn't the amount of power you'll be sending to your house, however. Your solar array will only be able to reap around 12%–15% of this energy, and these figures hold fairly well for wind generators, too, though efficiency percentage is not a commonly used term with home-based wind turbines, owing to the fact that similarly rated machines may have vastly different sweep areas.

However you measure it, it takes a pretty stiff breeze—square meter for square meter—to rival the power of the sun; far more wind than is blowing around in most locations. Considering the psychological effect wind has on a lot of people, this is probably a good thing. But it also makes your decision to install a wind system more difficult since you might live in a fringe area, where there may or may not be enough wind to make the installation of a tower and turbine a successful venture.

If the average annual wind speed where you live is 10 mph (16 kph) or more, you can almost be assured of having enough wind to reap a useful bounty of power from the unsettled atmosphere. This is because, unlike a solar array, a wind generator's capacity to produce power is not limited to the hours between sunrise and sunset—it can produce power day or night, rain or shine.

The DOE maps list wind speed in power classes from 1 to 7. The upper limit of Class 1 winds approach 10 mph (16 kph), provided the

A home in Norman, Oklahoma installed a Bergey 10kW Excel-S wind turbine as a grid-intertie system in 1983. The guyed-lattice tower is 100 feet tall. *PHOTO: BERGEY WINDPOWER*

Doug's Detours
Big, heavy and slow
aren't qualities we
often look for in
modern society, but
they're great qualities
in a wind turbine.

turbine is mounted high enough above ground. If you live in a Class 1 area you should probably do some homework before opting for a wind system. Living in a Class 2 area, though more promising than Class 1, does not assure you of enough wind, especially if your site is in a valley, near the lee side of a hill, or surrounded by towering trees —unless you are able to raise your tower at least 30 feet (9.2 m) above the tallest nearby trees. By contrast, hilltops, coast lines and high plains make excellent sites for gathering wind.

We cannot over-emphasize the importance of tower height in designing a wind system. There really is a lot more wind up there, and it will invariably be steadier and less turbulent than the gusty, chaotic breezes that occur closer to terra firma. In fact, the DOE generally considers the wind power density at 50 meters (164 feet) to be double what it is at 10 meters (33 feet). The actual figures will vary over different types of terrain, of course, but it's still an eye-opening exercise in mathematics. You probably won't erect a 164-foot tower, but the DOE figures do make a point: height is good (just like voltage).

Although most modern wind turbines begin to spin—and thereby produce some amount of power—at 6 to 7 mph (the "cut-in" wind speed), they will not really begin to produce much in the way of usable power below 8 or 9 mph (13 kph). For battery-backup systems, this is enough wind to keep the batteries charged and help reduce your electric bill. If you instead opt for a direct grid-tied system, you'll want an annual average wind speed of at least 10 mph (16 kph).

Whisper-ing in a Blizzard

In March 2003, Colorado was blasted by a drought-busting three-day blizzard. The snow piled up more than 3 feet on the level, with drifts of over 6 feet, completely covering cars and pickups, and solar arrays. Grid power went down in many places, and those around us who had solar with battery-backup eventually had to resort to their gas generators to keep things in the comfort zone. We, on the other hand, were basking in a power surplus, thanks to the steady 30-mph wind driving our trusty old Whisper 1000. Though we used electricity with giddy abandon (what else do you do when you can't go outside?) our battery reserves never dropped below 90%.   — Rex Ewing

Turbines: A Quick Look at the Windy Beasts

When home-based wind turbines are discussed, the image conjured up in your mind is probably that of a horizontal-axis machine. Consisting of a propeller, a rotor, a generator, and usually a tail, these turbines resemble wingless aircraft with oversized propellers. Though they may all look somewhat similar from a distance, there's a lot of difference between turbines, and your success or failure as a wind farmer will largely depend on which one you choose. Different machines are designed for different types of wind. Generally, machines with large sweep areas, such as Abundant Renewable Energy's ARE 110 and Southwest Windpower's Whisper 200, are engineered to operate optimally in lighter winds. Other machines, including the Whisper 100, have shorter propeller blades and are designed to take the punishment meted out at hilltop locations and during severe storms. Still other turbines, such as the machines produced by Bergey Windpower and Proven Energy, can endure some really nasty weather and still perform well in light winds.

Comparing wind turbines apples-for-apples will take a little research. Different machines have different cut-in speeds and different rated wind speeds, which is the speed at which optimal performance is achieved—usually in the 22- to 29-mph (35– 46 kph) range. The table on pages 118– 119 lists the rated wind speeds, along with other pertinent data, for a few popular machines. It should be noted that there are many well-built turbines on the market that aren't listed in the table. We're not playing favorites; we just wanted to give you a good cross-section for the purpose of comparison.

Practically all turbines on the market today are 3-blade machines. The 3-blade design runs smoother than a 2-blade unit, and will be a

Bergey Windpower's 7.5 to 10 kW Excel turbine. *PHOTO: BERGEY WINDPOWER*

little more efficient at converting wind into watts. As a general rule, the blades on smaller or lighter-duty machines are made from polypropylene, while those on heavier machines are epoxy-coated wood or fiberglass. If damaging winds sweep across your site from time to time, you should avoid plastic blades on turbines of 1,000 watts or more. Trust me; this is the voice of experience talking.

A braking mechanism is also a handy feature, especially if you live in an area with ferocious gusts that could possibly damage your blades, or where ice storms might cover them with a layer of hoarfrost or ice that can throw the system out of balance. With a wind brake—either mechanical or dynamic (electrical)—you can stop the turbine from spinning and wait for the sun to melt the frost or ice.

While the initial shock to your pocketbook will obviously be greater for a larger turbine, the ratio of wattage gained for money spent will also be greater. A larger machine will also outlast one or more smaller ones, and a heavy, slow wind turbine will have a longer lifespan than a light, fast one. So, if you're going the way of the wind, buy as much as you can afford. *(For more information on individual machines and the companies that make them, see chapter 11.)*

Electric Wind Turbines versus Water-Pumpers

When you think "wind generator" do you see a multi-bladed water pump, as used to dot the Midwest plains? Why aren't big, wide multiple blades used on turbines that make electricity? Water pumpers need lots of start-up torque to get things moving, which the large blade area provides. But once spinning, those multiple blades get in each other's airstream. Decades of careful experimentation have shown that two or three skinny blades will extract the maximum amount of energy from an airstream. Electric generators have almost no start-up resistance. Thus, the design of modern turbines with the minimum of skinny blades.

PHOTO: LAVONNE EWING

What Type of Power Do Wind Turbines Produce?

By the nature of their design, most wind turbines initially generate 3-phase alternating current by spinning three paired sets of magnets around three or more coils of wire. On turbines designed for battery-based systems, the AC is rectified into DC, either in the turbine itself, or within the charge controller. These machines will be configured at the factory for 12-, 24- or 48-volt operation.

Turbines designed for direct grid-tied operation, by contrast, churn out high-voltage DC from the turbine (up to 600 volts) that is converted to grid-compatible AC within the inverter. For this reason, direct grid-tied wind systems are more efficient than battery-based systems—up to 94% efficient—since there is no energetically expensive transformation from DC electrical energy, to chemical energy within a battery, and back again into electrical energy that must be further converted from DC to AC by the inverter.

Towers: Holding Your Turbine Up in the Breeze

Once you have a pretty good idea what size and type of wind turbine will fit your needs, you'll have to figure out how you're going to hold it up in the path of the wind. There are four basic types of towers used by most homeowners: guyed-pipe, guyed-lattice, free-standing lattice, and tubular monopole towers.

Proven WT2500 on an 84-foot guyed-pipe tower in Nevada. *PHOTO: SOLAR WIND WORKS*

Pipe towers are the cheapest, easiest to set up, and probably the most widely used. Made from sections of standard, off-the-shelf galvanized steel tubing, they are sleek, slim, and as inconspicuous as a tower can be—which isn't very. They are hinged at the base and then erected with the turbine already installed, blades and all. The major drawback to a pipe tower is that you cannot climb it for periodic inspections; it must be lowered.

Guyed-lattice towers are like the old ham radio towers. They are 3-sided and of uniform dimension from top to bottom—mine is around 18 inches (46 cm) per side—and, like pipe towers, must be supported by a series of guy wires. They can be assembled either vertically by sections, or on the ground on a hinged base and tilted up into place.

Free-standing lattice towers are broad at the base and taper toward the top, much in the same elegant way as the Eiffel Tower. Though more expensive (and showy) than pipe or guyed-lattice towers, you won't have to worry about clothes-lining yourself on a guy wire whenever you walk near it. Like the guyed-lattice towers, free-standing towers can be built in place, or assembled flat and raised to their vertical position.

A fourth type of free-standing tower, currently being offered by some turbine manufacturers, is the tubular monopole tower. These are like the solid, tapered steel towers you see holding communication equipment or lights high in the air above highway exit ramps. They're expensive and require a crane to erect, but are solid and good-looking, and they take up very little ground space.

Four Tower Types

Guyed Pipe *(page 73)*

Guyed Lattice

Free-standing Lattice *(above)*

Tubular Monopole *(right)*

TOP: BERGEY WINDPOWER
RIGHT: SOUTHWEST WINDPOWER

Most turbine manufacturers offer tower kits sized for each of their turbines, and those who don't sell the kits will make recommendations on which towers will work best with a particular turbine. Listen to these folks—they know what it takes to hold their machines up in the wind. The lateral thrust put on a turbine in a high wind is mind boggling, and nothing you want to experiment with. You wouldn't put a V-8 in a go-cart, would you? Same difference.

But even with good engineering, your tower and its foundation, like any structure, may be subject to regulation by your local building department. This means that you will have to comply with whatever codes are in place, since, unlike an un-permitted workshop tucked inconspicuously away in the trees, it's a bit difficult to erect a tower without anyone noticing.

In any event, there's no substitute for sound engineering, so unless you're an engineer and rigger by trade, you should seek professional assistance to ensure that all goes smoothly.

Which System Is Right For Me?

Southwest Windpower's Skystream 3.7 next to a suburban solar home.
PHOTO: SOUTHWEST WINDPOWER

What's true for solar is true for wind. If you live in the backwoods of Minnesota where blizzards and ice storms can tear down power lines in the blink of an eye and outages last for days on end, then you'd best be looking for a warm, cozy place to keep your batteries. On the other hand, if you live on the outskirts of a town with reliable grid power, then it might make more sense to opt for the more efficient direct grid-tied system. Component-wise, there is very little difference between solar and wind systems. Wind/battery systems are set up just like solar/battery systems. In fact, many wind charge controllers

The Power in Wind?

There's more than one way to skin a cat, as the expression goes, and there are various methods used to calculate the power density of the wind, each yielding different results. The common formula — wind power density $(W/m^2) = 0.05472 \times$ wind speed (mph^3) — is a QND (quick & dirty) way to an acceptable result. For the more painstaking methods used by NREL scientists for DOE wind studies, read the Wind Energy Resource Atlas of the United States at *http://rredc.nrel.gov/wind/pubs/atlas/chp1.html*.

have additional inputs for a solar array. And for direct grid-tied wind systems, the Windy Boy inverters from SMA America have the same features as the Sunny Boy and other comparable grid-tied solar inverters.

The main difference between solar and wind systems—beyond the obvious, of course—is that wind systems require more thought and more homework, owing to the plethora of different turbines and towers to choose from, as well as the widely varying wind conditions from site to site. Talk directly with the manufacturers, not just the guy who wants to install the type of wind turbine he just happens to sell. Make the manufacturer convince you that their machine is the one best suited to your needs. Do any of your neighbors have winds systems? Knock on a few doors and deluge the people behind them with questions; they'll be happy to display their vast knowledge about harvesting wind energy. In the end, the more you know before you buy, the more pleased you'll be with the results. ❖

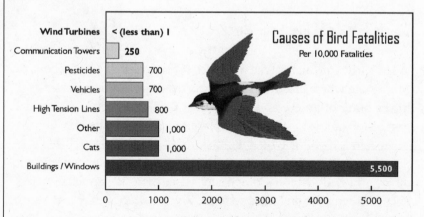

Source: "Summary of Anthropogenic Causes of Bird Mortality," by Erickson et al., 2002. (www.awea.org)

CHAPTER EIGHT

Let's Talk Money and System Performance

What Will a Solar / Wind Electric System Cost?

A solar / wind energy system large enough to take a substantial bite out of your electric bill is expensive up front. The average grid-tie system installed and permitted goes for around $20,000 to $30,000 before rebates or other incentives. Unlike most conventional power sources, solar's expense is all up front. There are almost no operational costs, it doesn't wear out no matter how hard you use it, and there are no long-term societal costs such as air pollution, environmental degradation, or reduced resources for our children. It's like buying years of electrical output in advance. All great reasons to feel warm and fuzzy, but it still needs to be affordable today.

We'll base these rule-of-thumb estimates on kilowatt-hours per day. Look at your monthly utility bill. Divide your monthly kWh total use by the number of billing days, usually about 30, to see about how many kWh you're burning per day. The U.S. average is around 20 kWh/day. What would you like to shave yours down to? Remember that your solar system doesn't need to cover all your electric use. Want to cut your electric bill in half? Can do.

What's it going to cost in your case? Based on the actual installed costs we've seen for direct-intertie systems over the past

Federal Tax Credits

Federal legislation passed in 2008 and 2009 includes a number of provisions for energy-conscious homeowners, including tax credits of 30% for new PV and wind systems installed between 2009 and 2016. For the complete details, visit: **www.dsireusa.org**

A Rough Estimate

For a direct-intertie system (without batteries), you can roughly figure investing $2,500 – $3,000 for every kilowatt-hour per day you want to shave off your utility bill.

few years, you can roughly figure on investing about $2,500 – $3,000 for every kilowatt-hour per day you want to shave off your bill. This is total cost, including installation and permits, but before any rebates, tax credits, or other incentives that will reduce the cost. Battery-based systems span a wider range of costs due to the variable size and quality of the battery bank, but figuring $5,000 per kilowatt-hour per day you want to save is firmly within the ballpark.

Grid-tie systems rarely try to cover your entire electric use. Sure, they could, but that's liable to be frighteningly expensive, and in most utility districts if you make more power over a yearly period than you use, you give the surplus to your friendly local utility company at no charge. Ouch! Nobody likes that prospect. However, most utilities let you carry over a credit from one month to the next within that yearly period. So if you make more power in one month than you use—a likely scenario in summer, or if you're on vacation,

Want to Start Small, with Room to Grow?

Grid-tied solar systems haven't favored those of us with modest budgets. The minimal start-up tab seems to be at least $10,000. The Enphase micro-inverter has changed that. With this innovative unit you can start as small as a single PV module, and you can add more PV wattage with one module at a time, or with many…your choice. The Enphase is a one-inverter-per-PV-module system. The initial startup is a little more expensive because you need the Energy Management Unit and some other bits for system monitoring (web-based preferred, but not required).

An approximate price for a minimal startup package with web-based monitoring for 5 years (a Sharp 175-watt PV module, flush roof mounting hardware, Enphase micro-inverter and miscellaneous hardware, Enphase Energy Management Unit 5-year subscription) would be about $2,500. Add another PV module / inverter next year and your cost would be approximately $1,500.

Please note we haven't included labor costs with these systems. Are we encouraging self-installation? Possibly. Because the Enphase systems don't involve extremely high-voltage DC there's less potential danger. But proper wiring and protection of 240-volt AC circuits is required, as is climbing and working on a roof. Sound daunting? Why not hire a pro and be his helper.

or this is a seasonal home—then you can use that credit later in the autumn or winter.

A direct-tie system will shave kilowatt-hours off the top of your bill. So those more expensive watt-hours over baseline rates are the first to go. Most homeowners who want to purchase a utility grid-tie system will size their system to deliver about 35% to 65% of their power use. These systems are very adaptable, because anything the solar doesn't cover gets picked up seamlessly by your existing utility.

Real-World Output (What should I really expect?)

Solar sales people will be happy to throw all kinds of confusing numbers at you, such as array wattage, inverter kilowatts, hours of daily sunlight, percent of full sun, etc. Yikes! Want it easy and honest? Here's a simple formula that delivers a fairly accurate, although slightly conservative, yearly average output (this assumes a due-south orientation with no shading for a direct-tie system):

(PV Array Wattage) x (Ave. Hours of Sun) x 70% = Daily Watt-Hours

Let's assume we have 18 Sharp 170-watt PV modules connected to an SMA Sunny Boy 3000 inverter. This is a fairly average-size system. **PV Array Wattage** is the manufacturer's label rating: in this case, 170 watts times 18 modules = 3,060 watts.

Average Hours of Sun is a yearly average of noontime-equivalent hours. This ranges from 4.0 to 5.5 hours for most of the United States. The National Renewable Energy Lab monitored hundreds of sites

How Big?

Most homeowners who want to purchase a utility grid-tie system will size their system to deliver about 35% to 65% of their power use.

A crew installs 4.7 kW of SunPower PV panels on the garage of a Maryland home.
PHOTO: NEVILLE WILLIAMS, STANDARD SOLAR.

Is It Going To Get Cheaper If I Wait?

Thanks to mass production, improved installation techniques, and smarter installers, the cost of solar has been gradually coming down. PV is more than 50% cheaper now than it was 10 years ago. On the flip side, every rebate program has tended to be better funded and more generous initially. As participation grows, rebates are usually diluted to cover more systems. The decision is yours.

from 1960 thru 1990, and then distilled it down very nicely in their Redbook. You can download individual states or the whole thing in pdf format at: *http://rredc.nrel. gov/solar/pubs/redbook/*.

The Sacramento, California page tells us that a south-facing, fixed array mounted at latitude angle (38°), or at latitude minus 15° (38°–15°= 23°) will receive 5.5 hours of noontime equivalent sun every day on a yearly average. Twenty-three degrees just happens to be about the same angle as typical 4:12 or 5:12 suburban roofs.

A fudge factor of **70%** takes into account all the real-world effects of dirty modules, dirty air, high humidity, hot modules, wiring losses, small bits of shading, inverter inefficiency, and all the other little things that are less than laboratory perfect in a working system. For battery-based systems, a factor of 65% is closer to reality. If your array is perfectly shade-free, installed perfectly, and you're in a dry climate above 6,000 feet elevation (and your karma is really good), then bump the factor up an extra 3% – 4%.

Angled Sharp solar panels fit nicely on this roof. *PHOTO: SHARP*

So our example will be: 3,060 watts x 5.5 hours x 0.70 = 11,781 watt-hours or 11.78 kWh per day. Remember, this is a yearly average. Most folks will see about twice as much power production in the summer as in the winter.

So long as there are no serious shading or other performance-reducing site problems, this formula usually yields a conservative estimate. Most customers report slightly better performance with their actual installed systems. "Better to look like a hero than a bum" has always been my motto, and you certainly can't depend on the weather to be the same every year, even in California. The long term Redbook charts state, "Uncertainty ± 9%." Take your result as a very educated guesstimate—your mileage may vary—but probably not more than 5%.

A home in New Jersey retrofitted with 2.7 kW of Evergreen solar panels.
PHOTO: EVERGREEN SOLAR

System Monitoring

How do you know how well, or poorly, your shiny new PV system is working? Where do you find bragging rights when your brother-in-law starts telling you about his bargain-basement grid-tie system that he paid some neighbor kid $50 to install for him? Ah! That's the system monitor, and there are several choices over how this data is delivered to you. We'll present them below starting with the simplest and working toward the more complex and expensive.

Inverter Display Every grid-tie inverter comes with an onboard display that will show how many watts are being output at the moment, how many watt hours have been delivered so far today, how many kilowatt hours have been output since first being turned on, and if there are any problems, it will display the fault code(s). Plus, there are

Monitoring Your Solar Array

There are many options for keeping tabs on your PV array's performance, beyond the readout on the inverter. Hardwired or wireless remote displays are handy, as is web-based monitoring.

Sunny Beam, a remote display. PHOTO: SMA SOLAR TECHNOLOGY AG

probably a few more readings the engineers threw in the rotating display. For many folks, this is sufficient. The display comes with the inverter, and all it costs you is the effort to walk out to the inverter and look. But sometimes that's difficult or impossible or just isn't as cool, which is why we have more choices.

Remote (Hardwired) Display   Most grid-tie inverter manufacturers offer a communications port that will send data to a PC computer within 300 feet (91 m). This comm port is standard equipment with some manufacturers, others charge up to $150 for the extra hardware. The software to read, display, and manage the data is universally a free download from all manufacturers. The CAT5E (Ethernet) cable to get from your inverter(s) to your PC can be simple or complex depending on the distance and tidiness required. The primary disadvantage of this monitoring type is that the PC must be turned on to receive the data. The inverter doesn't save up data till the computer is ready, it simply transmits data as it happens. If your PC isn't ready, well that's not the inverter's worry. If you've got a semi-retired laptop (laptop recommended due to much lower idle power use), data-catching could be a good use for it. Just pull the saved data off the laptop periodically.

Remote (Wireless) Display   Fronius and Sunny Boy both offer a wireless version of a remote display. These are separate stand-alone displays that don't require a home computer. Cost is in the $400 – $500 range. There's a small transmitter and antenna on the inverter. The receiver must be within about 150 feet (46 m), and if there's concrete, rebar, brick, or rock between, that distance can be much less. For the most part, these wireless displays will show the same info as the inverter-based display, but they can receive and display data from several separate remote inverters. The Fronius remote will cover up to 16 inverters, the Sunny Boy up to 12. In the case of the Fronius, the remote display IS physically the same display as on the inverter, it's just set up as a portable remote with no data storage. Sunny Boy's Sunny Beam is a cute little gizmo you can leave on the coffee table (bragging rights!) with several data windows available. It will store about 90 days worth of data and has a USB port for downloading.

Web-Based Monitoring We love our gizmos, and we love our remote access. Inverter and after-market manufacturers are hip to that, and are giving us more and simpler options. Want to see how well the PV system on your vacation house is doing? Want to show off your green creed to your fellow employees? Want to know if the battery-based repeater on some remote mountaintop is still working after the last storm? Web-based monitoring makes it easy. Every major inverter maker offers some way to plug their unit into your internet router, and several offer free or very low-cost sites to host, display, and store that data for you.

An Enphase monitoring window shows some early morning shading, with the darker-colored blocks indicating lower power output. *PHOTO: ENPHASE ENERGY*

Enphase, the new micro-inverter make (*www.enphaseenergy.com*) has their whole monitoring system based on web access. Their graphics-based displays are positively dazzling. In addition, there are a growing number of after-market companies that think they can do a better job of providing these services than the original-equipment manufacturers, and that you'll be willing to pay for the higher quality service (so far they're right).

Fat Spaniel (*www.fatspaniel.com*) is the first and most advanced of these companies with residential, commercial, and battery-based solutions. Most of these premium monitoring sites feature some automated analysis tools that will detect problems and send email or text warning messages. These are being widely used by better solar dealers as a service/maintenance perk. Your dealer will know

if your system is under-performing before you will, and can possibly provide repair services before you're even aware there's a problem.

Prices vary a great deal for web-based monitoring packages. There's a wealth of options and add-ons, but generally we're looking at adding about $1,000 or more to your system cost. It's not hard to add weather, wind, PV temperature, building power use, and other options into a system that may top $5,000. The danger of smorgasbords...

Are web-based monitoring systems the future? Absolutely! Here's my own site from GreenMeter (*www.sunsei.com/Monitoring*), an after-market monitoring provider for battery-based systems: *http://tiny url.com/c7y5nk*. It delivers a daily performance report via email and provides an easy way for me to check system performance while I'm on the road or away at work. It keeps a record of daily performance for the past year. Is it really necessary? No. The GreenMeter also feeds info to my home PC. Is it worth the extra money to put my data on a web-based service? For me, absolutely yes! The convenience and bragging rights make the one-time investment well worthwhile. ❖

If you're considering a standing seam metal roof for durability, fire resistance, or just plain good looks, then Uni-Solar's peel-&-stick laminate is the fastest, cheapest choice for adding solar power. This condo has an approximate 3,500-watt array (shown as the darker area of the roof). *PHOTO: UNI-SOLAR*

CHAPTER NINE

Any Financial Help Out There?

Yes, solar is expensive initially, but help is on the way. There are a wide variety of rebates, grants, buydown programs, tax credits, and exemptions available. But there are no blanket rules. Every state is different. The amount of help you'll be able to round up to pay for your particular system varies regionally. Some of these programs will cover 50% or even more of a complete system installed cost. There is, finally, thanks to the American Recovery and Reinvestment Act of 2009, a nice 30% federal tax credit with no cap for residential solar and wind electric systems and solar hot water systems that's available through 2016, and there are a wide variety of individual state and sometimes individual city or utility companies that are offering assistance.

The best place to start looking is at the Database of State Incentives for Renewable Energy (DSIRE) at *www.dsireusa.org*. Established in 1995, DSIRE is an ongoing project of the Interstate Renewable Energy Council (IREC), funded by the U.S. Department of Energy and managed by the North Carolina Solar Center with support from the University of North Carolina. DSIRE has a vast wealth of well-organized information. It's the best source to check first. Their info is accurate, and updated weekly. They'll give you bullet points, a written summary, and access info for each program. There isn't much that escapes their notice, except perhaps small,

Financial Help?

The best place to start looking is the Database of State Incentives for Renewable Energy **www.dsireusa.org**

local programs. There are a few small city-based programs it doesn't list, but that brings us to your next level of potential financing help.

Local Sources   Your second source of help will be local installers. These are the folks who will know about local assistance programs. Anyone who's working locally doing any kind of solar energy installations will be well-aware of any financial incentives that are available to their potential customers, and will be downright eager for you to hear about them. Start by looking in your local Yellow Pages under *SOLAR*. Next, in our appendix you'll find a full listing of the state energy offices for every U.S. state. Very often they will have listings of businesses offering solar services and installations. The amount of help your state energy office can offer will vary greatly, depending on whether your state is actively promoting renewable energy or not. In California you'll find lots of solar help; in North Dakota you'll do better to ask about wind energy.

A fully solar-fitted home with black pool-heating panels, a large PV array, and a domestic hot water collector at the far end, this homeowner is set to enjoy a lifetime of low energy bills.
PHOTO: SHELL SOLAR

If you're fortunate enough to be in a state with an active rebate program you'll probably find you've got a choice of local installers. Lucky you. Shop around. Prices and services vary widely. Ask dealers for local job references and then contact them. Any reputable dealer should have a string of happy customers they're pleased to share with you. Many installation companies are willing to do some or all of the paper wrangling for rebates, permits, permissions, etc. in return for your business. This is no small favor. The learning curve on these forms is steep, and the time required isn't slight. Knowing how to deal effectively with the many bureaucracies involved can save you major frustrations and delays. If you have a licensed installer who's also willing to handle the paperwork, count yourself blessed! Some companies may even offer to accept the state rebate in lieu of payment from you. Those are dollars that never need to leave your pocket, and since the installer can't claim the rebate until your installation is finished and signed off, you're assured of a speedy, code-compliant install.

Return on Your Investment

Installing a solar electric system increases the value of your home, particularly as Americans continue to become more energy conscious. According to a now somewhat dated, but extensively researched study from the National Appraisal Institute (*Appraisal Journal,* Oct. 1999), your home's value increases $20 for every $1 reduction in annual utility bills. This is a ratio that will only increase as energy costs rise. At 1999 rates, a modest solar-electric system of 2,500 watts will increase your home's value by $8,000 – $10,000 immediately, just for the utility bill reduction. Many states have adopted property tax exemption laws, such that if your home value is increased by a solar installation, you cannot be taxed on that increased value. You'll find details at the DSIRE website.

Renewable Energy Credits or Green Tags

If your solar or wind electric system is providing renewably generated energy, not only do you get an immediate reduction in your utility bill, but you are creating a commodity called Renewable Energy Credits, or Green Tags. These represent the environmental benefits of your clean energy separated, or unbundled, from the actual kilowatt-hours of energy. In most states these credits can be sold or traded like any commodity. Essentially, you're selling the bragging rights to your green power, but not the actual power.

    It works like this: say some major corporation wants to paint a

Whose Green Tags are These?

Green Tags are a commodity that can be traded and sold (see next page), but there's controversy over who owns them. California utility giants are claiming they own the tags; most other states have laws that specifically give the tags to the homeowner. Until the courts decide, California homeowners are out of luck.

Looks Do Matter

Installations that are high quality and good-looking do have a positive impact on a home's resale value. Get it done right, from PV panels to wiring to placement of components.

greener image of themselves (no really, it happens sometimes!). Rather than invest big bucks in their own solar electric system, Mega-Bucks Inc. buys up the green tags from a number of smaller systems that a broker has assembled. Now Mega-Bucks can legally say, "We get X percentage of our energy from renewable sources, we are soooooo good you should just run out right now and buy all our products!" Once you've sold your green tags, you can no longer say that you get your energy from a renewable source, even though you certainly continue to in reality. You can say that you're hosting a solar energy array, and you can point to it and say, "See the nifty solar array on my house? It makes an average of 20 kWh per day!" You just can't claim the environmental benefits. But then you've got some extra bucks in your pocket that you can claim. Credits vary depending mostly on the size of your system. Most residential systems are currently worth $50 to $500 per year for the green tags, and you can continue to sell your credits (or not) on a yearly basis. ❖

The chimney, and the shade it casts in the afternoon, cost space for three modules on this rooftop. Since it didn't seem likely they'd move the chimney, the system designer wisely found a solution and installed 17 modules (2,975 watts)—very nice-sized array for moderately conservative homeowners.

PHOTO: AFFINITY ENERGY

CHAPTER TEN

Permits and Paperwork

Paperwork. It's nobody's favorite thing to deal with, but we've got some tips, suggestions, and hard-won experience that can make it less of an ordeal.

What to Expect

Because we're writing this book for homeowners from Seattle to Miami, we can only generalize about the particular forms you may need for your installation, but we've found there are at least three forms you can count on:

Luminous Abundance

The earth receives more energy from the sun in just one hour than the world uses in a whole year.

SOURCE: SOLARBUZZ.COM

1. State or local government programs are going to have some forms to complete if you expect to cash-in on a rebate, grant, buy-down, or however they've structured it. In most cases you need to be pre-approved, so this is generally the first form to file. After approval, your funds are put in escrow, and you've got a reasonable amount of time to install the system, typically 6 to 12 months, before you come back to claim your dollars.

2. A local building permit from your city or county is a necessity, and typically the rebate folks will want a copy

Keep Your Insurance Up-To-Date

Don't forget to tell your homeowner's insurance company about your new green energy additions.

of this when it's signed off as proof that the installation actually happened, and it was done to comply with building and electric codes.

3. In order to correctly setup billing and credit, and record your home on their distributed-generation maps, the local utility company will have their own set of forms, and sometimes an inspection. They usually require a signed-off building permit submitted with their forms, so this is probably the last process that needs attention.

Find Somebody Else To Do It!

For those who really hate paperwork, our first tip will save a few gray hairs. Let somebody else do all the paper wrangling! This isn't as far-fetched as it might sound. If you're in a state with a good, active rebate program there are likely to be several contractors or installation firms that want your business. One of the ways they can get it is by handling most or all of the paper hassles. As someone who has been in this business since the beginning and has wrangled more than his share of rebate paperwork, take my word...this is a great incentive when you can find it! Even for pros who know exactly what's required by which bureaucracy, this service probably represents several hours of someone's time in the background. And that's if Murphy's Law doesn't rear its head, and everything goes smoothly without "lost" paperwork or misunderstandings. Be prepared to pay a bit extra for this kind of service, and then feel good and smug about it.

A Gaithersburg, Maryland home with a 2.7 kW grid-tied system of Isofoton PV modules.
PHOTO: NEVILLE WILLIAMS, STANDARD SOLAR

Do It Yourself?

Feel like you can handle this yourself? Good for you! But first a warning about doing the physical hardware installation yourself. Most of the big rebate programs have a self-installed penalty clause. In California and New Jersey it's a 15% reduction in your available rebate. They'd rather not have contractors and industry insiders using up all the funds, for obvious reasons. So even if you are a contractor, and you really are buying this system for your own house, you might want to hire a fellow contractor to sell and install it for you. With the extra 15% rebate you can afford it, and it'll probably ease the paperwork hassles a bit.

A Columbia, Maryland home outfitted with a 3.15 kW array of BP Solar panels.
PHOTO: NEVILLE WILLIAMS, STANDARD SOLAR

Now, as to doing the paperwork yourself, all the major programs have forms and guide books available online. Just download and go for it. Be sure to fill in ALL the info. If you aren't sure what they want, call them. These programs are heavily subscribed to, and it sometimes takes 3 or 4 months for them to process your paperwork. To have forms returned because of missing info—a few months later—is discouraging, to say the least.

Don't fax, FedEx. Another tip born of experience. Although most of these agencies give you a fax number, we've found that the occasional unreadable item—that doesn't get noticed and returned until 3 months later—makes the extra $20 for express delivering the originals seem like a bargain. FedEx or UPS also gives you proof-positive of precisely when an application arrived. Handy proof to have when an application gets "lost."

Keep copies of everything! This is very important, and has pulled my bacon out of the fire more than a few times. The agencies doing solar rebates are big, busy, oversubscribed, and understaffed. Stuff gets lost, misrouted—who knows? Having copies of everything you submitted is a must. Hopefully you'll never need them. But when you do need them, you'll need them BAD! In a worst case this can save you from having to start all over at the end of the line.

Typical installation with a Sunny Boy 2500 inverter in the center (cover off for wiring), a DC disconnect on the right, and an AC disconnect on the left. Far left is the existing house meter and main breaker panel, also with the cover removed for ease of wiring. *PHOTO: DOUG PRATT*

Tell Your Friendly Local Utility Company

Your grid-tie system will function perfectly whether your utility company knows about it and has given their official blessing or not. Hardware is funny like that. In fact, your installer should turn it on briefly just to check that everything is functioning correctly at the end of the installation. But briefly to check performance is all that it should be on until the utility company does their inspection (if required) and gives permission to start intertie. The utility companies take unpermitted power input very seriously. Get caught pushing your meter backwards without permission and you can very easily find yourself locked off from grid power and facing some stiff fines.

So fill out the utility company paperwork. Besides earning the right to spin your meter backwards legally, this will get you on their local mapping so that repair crews know you're a potential generator. It also ensures that you've got the correct type of meter installed. The standard clockwork type meters will spin forward or backward just fine, but the one you've got now was only certified to be accurate going forward. They'll probably install a new bi-directional meter for you. You may have some choices about what kind of billing program you're going to be on. If your house is usually empty during the day, you may want to consider Time-of-Use billing, if it's available for you (read on).

Buy Low, Sell High

That old stock market saw has a new meaning in the California utility intertie market. The state legislature has ruled that any billing scheme normally available to a homeowner has to continue to be available to them if they install a solar electric system. This means that Time-of-Use (TOU) billing is available to solar grid-tie homeowners. The idea behind TOU was to reduce electric demand on weekday summer afternoons by raising rates between 1:00 p.m. and 7:00 p.m. Now solar homeowners can sell kilowatt hours for about 30 cents in the afternoon, and buy them back for about 15 cents in the evening. It's legal, and the credit shows up as dollars on your bill. This has naturally become very popular with grid-tie system owners, because it will speed your payback by 15%–25% for most households. Time-of-Use favors the typical modern household where everyone's off to work or school most of the afternoon. You want to be sure to schedule electric chores before 1:00 p.m. or after 7:00 p.m., as much as practical. Pool pumps and air conditioning are good candidates for rescheduling. TOU billing has a summer schedule, May through October, when there's a large differential between peak and off-peak rates, which just happens to largely coincide with peak solar output in most locations. There's a much smaller rate offset the rest of the year, currently about 10 cents and 8 cents, but you still get to sell in the afternoon at a slightly higher rate than you buy back in the evening.

Going Solar In A Big Way

In 2008, a record-breaking 342 megawatts of PV was installed throughout the United States. California led the way with 178.6 MW in grid-tied installations, followed by New Jersey (22.5 MW) and Colorado (21.6 MW).

SOURCE: WWW.RENEWABLE ENERGYWORLD.COM

Several ground-mounted arrays (15 kW of solar power) supply a grid-tied home outside Albany, New York. *PHOTO: GROSOLAR*

U.S. Wind Power

At the end of September 2008, approximately 21,017 megawatts of utility-scale wind installations were operating in 35 states across the U.S., plus another 8,000 megawatts either planned or under construction. This does not include the tens of thousands of small wind turbines used directly by homeowners.

SOURCE: AMERICAN WIND ENERGY ASSOCIATION

Now Comes the Good Part...

Once your utility representative has inspected and given their blessing, you're free to let those high-powered solar-generated electrons rip. Turn the system on and enjoy watching that meter spin backwards. If you make more energy than you use in a month you'll probably find a credit on your bill. Every state, and sometimes every utility company within a state, has their own rules for how to deal with credits. The most common scenario allows you to roll credits forward from month to month for a 12-month period. If you end the year with credit still on the books you'll probably give it away, although some utilities will pay you their wholesale price, and others simply cash you out at wholesale rates every month. Make sure you understand how your utility deals with credits, because you can adjust your lifestyle to some degree. Invite all your friends with electric vehicles to come over for instance, or crank up your hydrogen electrolyzer, or I've even heard stories about some kinds of agricultural pursuits and grow lights. If you've got extra, you'll find a creative way to use it.

Don't Get Too Smug About It

A small word of warning is needed at this point. I've known some folks with modest grid-tie systems that ended up with a higher electric bill after they turned their system on. No, it was working perfectly. The inverter's built-in cumulative watt-hour meter showed their system was actually delivering slightly more energy than predicted. It was a mental attitude thing. "Oh we've got solar power now, we don't need to worry about turning the lights off." In the energy conservation biz we called this "take back" and it's common. "Oh, I've got solar hot water, so I can hang out in the shower a few minutes longer without feeling guilty." We all do it; it's human nature. Just be aware you're doing it, and don't whine about the results. ❖

CHAPTER ELEVEN

The Nuts and Bolts: What To Look For, What To Avoid

Okay, you've decided that solar or wind energy seems like a good idea, and now you're shopping for the right bits and pieces for your system. This chapter is where you get to tap our many years of experience with good, bad, and indifferent solar equipment manufacturers. I've always believed that honesty was the best sales tool, and, so much as editors and lawyers allow, you'll find real-world honest assessments here.

Solar Modules

Close to 80% of your investment is probably going to go into PV modules, so your choice matters...or does it? The good news is that module quality is universally excellent. You should expect a product with a 20- to 25-year performance warranty, and regardless of manufacturer, you'll enjoy long, trouble-free ownership. The bad news is that solar energy has become immensely popular around the world, and supplies remain tight, especially in the summer months when most installations are done. So whatever brand and wattage of PV modules your dealer can actually deliver is probably a good brand. There's a rising flood of Chinese modules that seem fine so far, but with limited performance histories, I'm hedging a bit. Generally bigger is

Good News!

Your biggest investment in a solar electric system – the solar modules – have the best warranties... 20 to 25 years.

better. PV modules are sold by how many watts they deliver under a standardized test. With higher-wattage modules you don't need as many, which results in less bolting and wiring. Most power production modules are in the 150- to 250-watt range now. All that being said, there are a few other fine points to consider in module selection.

Single-Crystal vs. Polycrystalline vs. Hybrid Solar Modules

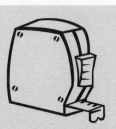

Single crystalline was the original construction technique, but polycrystal is easier to produce, and these days, is just as energy dense. If you need maximum wattage in the minimum space, and price isn't as important, look to the new hybrid modules. They're typically polycrystal cells with a thin, amorphous, secondary PV layer on top. Sanyo and SunPower are currently offering them, with more in the wings. A selection of various PV modules are detailed below:

BP SC3200 200-watt polycrystal	66.1" x 32.9" = 15.13 sq. ft.	(1.40 m²)	*13.2 watts/sq. ft.*
Evergreen ES-A 210-watt polycrystal	65.0" x 37.5" = 16.9 sq. ft.	(1.57 m²)	*12.4 watts/sq. ft.*
Kyocera KD210GX-LP 210-watt polycrystal	59.1" x 39.0" = 16.0 sq. ft.	(1.48 m²)	*13.1 watts/sq. ft.*
Sharp ND224U1F 224-watt polycrystal	64.5" x 39.1" = 17.5 sq. ft.	(1.63 m²)	*12.8 watts/sq. ft.*
Sharp NT175U1 175-watt single-crystal	62.0" x 32.5" = 14.0 sq. ft.	(1.30 m²)	*12.5 watts/sq. ft.*
Sanyo HIT 205-watt hybrid poly/amorphous	51.9" x 34.6" = 12.4 sq. ft.	(1.16 m²)	*16.5 watts/sq. ft.*
Uni-Solar PVL 144-watt amorphous	216.0" x 15.5" = 23.25 sq. ft.	(2.16 m²)	*6.2 watts/sq. ft.*

PV Connections

PV modules are equipped with either junction boxes (J-boxes), each with a terminal strip inside, or more commonly, with multi-contact cables (MC cables), which have become the preferred connector for direct grid-tied systems, where we need to make a lot of series connections. The MC connectors just plug together in series, making module wiring very fast and fool proof. J-box modules are usually preferred for battery-based systems, which use fewer or sometimes no series connections at all.

In either case, don't pass up a batch of available modules just because they don't have the optimum connector type. This is just a matter of convenience. Either connector type will function okay with any installation. In the worst case, you or your installer might spend an extra couple hours on wiring.

Second-generation PV connectors are required by the 2008 National Electric Code. These connectors must physically latch so they can't be accidentally pulled apart. And if they're installed in an "accessible" location, like a ground mount, there must be additional locking connectors that require a tool to separate them. Double-insulated cabling is also starting to become the norm for PV manufacturers. This higher quality cabling allows the use of transformerless inverters. Transformerless inverters are about 1%–2% more efficient at turning PV power from your modules into AC power for your house. So this is a good thing.

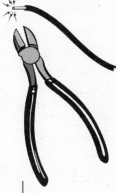

Be sure that when the installation is finished, any extra MC cabling is secured neatly under the modules and isn't laying on the roof surface. Twenty years of the wind blowing a cable back and forth across an asphalt shingle will wear it down to nothing, exposing the high-voltage wiring. Quality installations use stainless or galvanized spring steel clips to secure the cabling.

When you're hanging PV modules over the front door of your million-dollar home, aesthetics are important! This 7.0 kW black-framed Sharp array with triangles fits the home beautifully.
PHOTO: SHARP

Hybrid Solar Panels

If you need maximum wattage in minimum space, and price isn't as important, look to the new hybrid modules. They're typically poly-crystal cells with a thin, amorphous secondary PV layer on top.

Module Construction

Single-crystal and polycrystalline modules with full-perimeter aluminum frames and tempered glass covers are the norm, and will deliver the most wattage with the least surface area. Amorphous silicon modules require about 50% more surface area to collect the same amount of wattage, but can be built without glass covers for an unbreakable module. Uni-Solar® offers unique amorphous products designed to be the roof, not just sit on it. They have a peel-and-stick product for application on commercial membrane or standing-seam metal roof pans. (It's greatly preferred to do the application on new pans before they're installed.) These products are vandal-, theft- and hurricane-proof. Installed costs are similar to conventional modules.

Ingenious PV modules designed to interweave with conventional ceramic tiles are starting to appear on the market. Some are even available in earth-tone colors when aesthetics matter.

Uni-Solar's unique PVL thin-film solar modules are nearly invisible when applied to conventional charcoal-grey metal roofing (arrows point to the solar roof), as shown on this 1.5 kW Research Townhome by the National Association of Home Builders. *PHOTO: UNI-SOLAR*

Module Mounting Systems

Most mounting systems are made of extruded aluminum rails in various pre-cut lengths with slots top and bottom that permit adjustment for various module sizes and rafter spans. Tilt-up leg hardware is added as needed. Attachment to the roof is usually done with standoffs which penetrate the roofing and are waterproofed with a flashing just like a plumbing vent (which every roofer understands). With conventional asphalt shingles, the more expensive standoffs are optional. With shake or tile roofs, standoffs are required (*see pages 58 and 130*).

Ground-mounted PV arrays use the same rail mounting systems; we just add legs for small arrays, or an engineered pipe framework for larger modules (*also see pages 56, 57, and 93*). Poletop mounts hang the array from the top of a single steel pipe, usually 4- to 8-inches (10 – 20 cm) in diameter. These pole-mounted arrays are less popular for grid-tied systems because we can't usually hang more than about 10 modules on a single pole due to wind loading, and grid-tied systems often run 20 to 60 modules.

The back side of a pole-mounted solar array shows Unirac's mounting system and the black junction boxes of the solar panels. *PHOTO: UNIRAC*

For all types of installations you'll want to check out a new mounting system from Next Gen Energy (*www.ngeus.com*) called Zilla™. It is versatile, omni-laterally strong, and requires fewer roof penetrations which reduces labor time and costs. It can be used for PV panels as well as solar thermal installations. Other quality mounting systems are Unirac (*www.unirac.com*) and Direct Power and Water (*www.directpower.com*).

The Zilla mounting system's foolproof design and easy assembly simplifies installations. *PHOTO: NEXT GEN ENERGY*

Spanish, or barrel tile, is a special case. Just stay off it if there's anyplace else to put your array! It's difficult and expensive to work with. If the roof is truly the only place, then Professional Solar Products (*www.prosolar.com*) makes Tile Trac®, the best mounting hardware package for Spanish tile. It puts no weight on the tile, attaches securely to the rafter, and still manages to penetrate the tile at its highest point so there's little chance of leakage. It doesn't make tile mounting fast or fun, but does make it possible.

Installed on the Grove dance club, the mounting system holds 20 Kyocera panels, along with two PV Powered inverters.
PHOTO: SUNLIGHT SOLAR ENERGY

Increasingly, we're finding that new homes are using engineered trusses. When the truss engineer knows that it's going to be a tile roof, he does not add extra load-carrying capacity to the trusses (a second layer of tiles will never be added). So there's no load capacity available for a solar electric system. The only way to do solar on a roof like this is to take the tile off in the area where the array will be placed. That frees up plenty of carrying capacity. Put down a layer of rolled roofing or asphalt shingles, and then do your conventional PV install. Fit tiles in around the edges and run a row or two along the bottom below the array. Looks great and is still perfectly watertight.

Why Don't I See Tracking Mounts?

A tracking mount is one that follows the sun from east to west every day. By facing the modules directly at the sun, daily output can be boosted up to 30% in summer, or 10% – 15% in winter. Ten years ago trackers were common. Since then, the cost of PV modules has dropped 50%, while tracking mounts have only gotten more expensive. Plus, trackers tend to require regular maintenance, and finally, none of the current rebate programs give extra credit for the expensive tracking mounts. Economics simply don't favor trackers any more. Want more output? Buy another module or two, put them on a fixed mount and never think about it again. There's much to be said for *no moving parts!* Except for some specialized pumping systems the era of tracking mounts has pretty much passed.

Direct Grid-Tie Inverters

Here you've got some choices! There's a wealth of good inverters on the market now (*the most popular models are listed alphabetically below*) with more showing up every few months, and there's no lack of inventory. Your dealer should be able to get you any model you want with no problem. Most residential inverters will come with a 10-year manufacturer's warranty, because the industry-leading California rebate program requires it. If 10 years isn't standard, it will be available as an option. In early 2005, the California Energy Commission (CEC) started requiring independent third-party evaluations of inverter efficiency. Since these ratings come close to reflecting reality, and affect rebate payments, we'll list CEC efficiency. The major inverter brands are listed below. The most popular brands, plus our personal favorites will get covered in more detail.

Enphase Micro-Inverters *www.enphaseenergy.com*

Enphase is the first company to produce a one-inverter-per-PV-module system. Each module has its own micro-inverter installed underneath. Output is conventional 240VAC. No string sizing, no restrictions on how few or how many PV modules are used. Do you have complex, unavoidable shade problems? Would you like to start out small and add a PV module or two to your array every year? The Enphase micro-inverter system is ideal for both of these situations. Because each PV module is a complete system, each module will deliver the maximum wattage it can under current conditions, regardless of what its neighbors are doing. The only disadvantage to this system is cost. Enphase inverters currently cost about 30% more than conventional string inverters, although the manufacturer is expecting that difference to diminish as production ramps up. Enphase inverters are made for a specific

An Enphase micro-inverter attaches to every PV module.
PHOTO: ENPHASE ENERGY

number of silicon cells per module, and for a specific wattage range for that module. So there isn't an Enphase model for every PV module on the market...at least not quite yet. The housing has a NEMA 6 (submersible) rating. CEC efficiency ranges from 94.5% to 95.0% for models released so far. Standard warranty is a very impressive 15 years.

Enphase monitoring is equally innovative. The individual inverters blip their performance data down the AC wiring along with the output power. It's picked up by the Energy Monitoring Unit which is plugged into any convenient outlet in your house. This info is sent via broadband (if available) to Enphase's web monitoring site where it's displayed graphically and stored for you. The Enphase web-based monitoring site is one of the slickest and most informative websites this old solar greybeard has ever seen. If there's no internet access on-site, the info can be sent to a conventional PC with web browser. One Energy Management Unit can handle up to 240 inverters.

Enphase
Energy Monitoring Unit.
PHOTO: ENPHASE ENERGY

NEMA? What's a NEMA?

NEMA (National Electrical Manufacturers Association) provides industry-standard ratings for electrical equipment, as in what sort of protection the housing on your electric gizmo offers, or what kind of environment it can be installed in. Higher numbers generally denote more protection. The most common ratings are:

◆ **NEMA 1**: the most basic rating to keep little fingers out; for indoor use only.

◆ **NEMA 2**: for indoor use only, but with added protection from dripping or light splashing of liquids.

◆ **NEMA 3**: for indoor or outdoor use; provides protection against rain, snow, sleet, falling dirt, or windblown dust.

◆ **NEMA 3R**: same as NEMA 3, but without protection from windblown dust. (Probably due to ventilation slots.)

◆ **NEMA 4**: same as NEMA 3, but with added protection from hose- or sprinkler-directed water streams. (No ventilation slots, probably watertight sealed.)

NEMA 4 is typically the highest rating we'll find on any household electric equipment. But just so you'll know, NEMA 4X adds corrosion protection (used in the marine industry), NEMA 6 adds brief submersion, and NEMA 7 is for explosive environments.

Fronius IG- and IG Plus-series Inverters

www.fronius.com/cps/rde/xchg/fronius_usa

Fronius is an Austrian company with roots in DC welding and battery-charging technology. Thanks to the German solar market they've been producing grid-tie inverters since 1995. The 2nd-generation IG Plus series features seven models from 3,000 to 11,400 watts—the largest single residential inverter currently available. Fronius has several unique features. Using small, powerful, high-frequency transformers, the weight is dramatically reduced, and adaptability is increased. Long-life fans are used at proportional speeds when extra cooling is required. Their 4,000-watt and larger units have multiple power channels. When incoming power is lower, only one channel operates, which gives better efficiency. The "master" channels alternate automatically every other day.

Fronius IG Plus inverter.
PHOTO: FRONIUS INTERNATIONAL GMBH

The new IG Plus series includes a built-in DC disconnect switch (UL-listed) and a six-circuit PV combiner, so those necessary parts don't clutter up your wall space. They have simple expansion slots, like a computer, for easily adding remote monitoring, environment sensing, or other future improvements. Their power boards are completely dipped in conformal coating for the greatest resistance to moisture and corrosion (most manufacturers just coat the back of the board). If you're in a humid environment, this could mean a much longer life expectancy. The NEMA 3R housing is okay for indoor or outdoor mounting.

The Fronius inverter also has a nice full-featured LCD display, easy mounting, quick wiring connections, and they're FCC compliant for greatly reduced radio/TV noise. Setting up an optional remote computer or web-based display is quick and simple with standard Ethernet connections, or a wireless remote display is available. Altogether, Fronius makes impressive inverters with an almost flawless track record. CEC efficiency ranges from 95% to 96%, depending on model. Standard warranty is 10 years.

PV Powered inverter.
PHOTO: PV POWERED

PV Powered Inverters   *www.pvpowered.com*

PV Powered is an Oregon-based company. (Yes, Oregon is part of the U.S. the last time anyone checked.) Designed on the KISS principle (*Keep It Simple Stupid*), PV Powered inverters have by far the fewest number of parts of any inverter on the market. Fewer parts means less chance of something failing, thus the expectation that this should be a very dependable, long-lived inverter. PV Powered currently offers seven models from 1,100 to 5,200 watts that are all FCC compliant for radio noise. Thanks to extremely high efficiency, an oversize heat sink, and good thermal design, PV Powered inverters run cooler, don't require a fan, and don't require any power limiting software features at higher temperatures. They simply won't get hot. The NEMA 3R enclosure is approved for indoor or outdoor mounting. An inverter-mounted fluorescent display is standard. Optional hardware is available for web-based or remote monitoring. CEC efficiency ranges from 90.5% to a chart-topping 96%, depending on model. Standard warranty is 10 years.

Inverters You Don't Want

Let's take a moment to mention some inverters you don't want. It's unlikely you'll run into either of these unless you're shopping for used equipment, but just in case...

AES (Advanced Energy Systems) GC-1000. Many of these small 1,000-watt inverters were installed until about 2002 when this U.S.-based AES went belly up. (There's a different AES out of Australia now; a completely different company.) They don't seem to have a particularly long-life expectancy and there's no repair or warranty service available. You don't want a GC-1000 unless it's free (and even then it's iffy).

Trace/Xantrex Sun-Tie Series This was the first mass-produced, direct-intertie inverter on the market. You may still find a few of these for sale cheap. They were rushed to market and suffered a wide variety of problems. Xantrex had just bought Trace, and while getting settled had some trouble admitting any faults. Eventually they genuinely tried hard to make good. By then the bad reputation had caught hold and it was like trying to sell return tickets for the Titanic's maiden voyage. They couldn't give them away. Better to stay away from this discontinued model.

SMA Sunny Boy and Windy Boy Inverters

www.sma-america.com

SMA of Germany makes Sunny Boy inverters, easily the most popular inverter brand in North America. They have been producing inverters in Germany for the European market since the mid-1990s, so there's considerable experience and a wide range of models to choose from. Sunny Boy America came out with an almost completely new product line in 2007, with five models ranging from 3,000 to 7,000 watts. They all feature built-in 4-circuit-string combiners and a DC disconnect. This makes for very fast, clean installations, with a minimum of wall clutter. All models feature powder-coated stainless steel enclosures for outdoor (preferred) or indoor mounting, and they have proportional cooling fans. SMA dropped their bright red color scheme; the new models are a plain, but universally acceptable beige color.

This well-done installation on a Palo Alto, California home includes two SMA 7000 inverters connected to a huge array of nearly 15,000 kW.
PHOTO: RENEWABLE POWER SOLUTIONS, INC.

The standard inverter-mounted display shows instant wattage, cumulative watt-hours for the day, cumulative since installation, and a few other readings, plus any error codes. Remote, web-based, or wireless remote monitoring options are available for single or multiple inverters. All SMA inverters are FCC-compliant for low radio noise. Sunny Boy inverters are competitively priced, with larger units being more cost effective. CEC efficiency is an excellent 94.5% – 96%, depending on model. Standard SMA warranty is 10 years.

At this time, SMA is the only source of UL1741 certified inverters for direct grid-tie wind turbines. Any of the new 3,000 to 7,000-series inverters can be configured as (what else?) a Windy Boy inverter. Since the Windy Boy software has to be custom-configured to work with each particular brand and size of wind turbine, you'll only find the Windy Boy offered as a package of turbine and inverter(s) together. Your turbine will require some additional control hardware to keep things in check should the utility fail during a windy period. (Power goes out when the wind blows? Gee, really?) See chapter 7 for more info.

Wind Turbines for the Windy Boy

Which turbine manufacturers produce models compatible with SMA America's Windy Boy inverter? Currently the Abundant Renewable Energy's ARE 110 and ARE 442 turbines can be configured to work with the Windy Boy, as well as the Kestrel e300i and e400i models, and all the Proven Energy machines. Southwest Windpower's Whisper 200 can likewise be adapted. Others will surely follow. Because of the widely varying electrical characteristics of different machines, the Windy Boy inverter must be configured at the factory for your specific model of wind turbine.

Solectria Renewables Inverters   *www.solren.com*

Solectria grid-tie inverter. *PHOTO: SOLECTRIA*

Solectria is an east coast U.S. company that got its start building robust inverters for electric vehicle motors back in the 1990s. You think household inverters have to be tough and reliable? They've got it easy compared to the vibration, water, mud, and diverse temperature range of automotive use! But you may have noticed we aren't all driving electric vehicles (or flying cars either) quite yet, so Solectria rolled their expertise into products they could sell now. They offer several well-regarded large industrial inverters, which we aren't going to cover here, and six excellent residential units from 1,800 to 5,300 watts. Their 3.0, 4.0, 5.0 and 5.3 kilowatt-units are the real standouts with integrated 3- or 4-string combiners, a built-in DC Disconnect, lightweight design, standard-equipment data ports, and industry-leading 96% CEC efficiency for all four models. They've managed to combine all the best design bits from every other inverter manufacturer into a single model line. Thanks to their rugged heritage, Solectria inverters have proven to be extremely reliable, unaffected by weather extremes, and are simple to install. They will continue at full-rated output under much hotter conditions than most inverters will tolerate (130°F or 55°C).

Onboard monitoring is provided by a full-featured LCD display. For remote or web-based monitoring, RS-232 and RS-485 ports are standard equipment and PC-based monitoring software is a free download. All Solectria inverters are FCC-compliant for low radio noise. Solectria residential inverters are competitively priced, with larger units being more cost effective. The 3.0 kW and larger Solectria inverters use a NEMA 3R enclosure that's okay for indoor or outdoor installations. The smaller 1,800- and 2,500-watt models feature a completely sealed NEMA 4 enclosure, and are CEC rated at 92.5%–93%, depending on model. Standard warranty for all Solectria residential inverters is 10 years.

Xantrex GT-series Inverters   *www.xantrex.com*

The GT-series are Xantrex's second-generation direct grid-tie inverters, engineered in the U.S. and Canada, and built in China. Four models are currently available at 2,800 to 5,000 watts. A built-in single-knob AC and DC disconnect is included. No fans are required for cooling.

A full-featured LCD display is included, as is Ethernet or RS-232 connections for computer monitoring. There's also a hard-wired remote monitor available, but no wireless or web-based monitoring (yet). It's FCC-compliant, for minimal radio noise or interference. The NEMA 3R enclosure is approved for indoor or outdoor installations. CEC efficiency ranges from 94% to 95.5%, and standard warranty is 10 years.

Xantrex's GT 2.8 grid-tie inverter. *PHOTO: XANTREX*

Battery-Based Grid-Tie Inverters

Battery-based systems haven't been as popular as direct-tie, but if your utility power is subject to interruptions due to hurricanes, ice storms, or Enron traders getting rich, then battery-backup is your ticket to independence. Your list of inverter choices is shorter and simpler, and CEC efficiency will be slightly lower due to battery charging losses. Here are the most popular options *(listed alphabetically).*

Beacon's inverter for battery-based systems.
PHOTO: BEACON POWER

Beacon Power   *www.beaconpower.com*
The Beacon Power M5 is a U.S.-made 5,000-watt inverter with an advanced 3-channel MPPT charge controller, switchgear, and ground fault protection all housed in a single NEMA 3R indoor/outdoor box. It offers instant 120VAC backup power and requires minimal space. Batteries, battery housing, and battery overcurrent protection are sold separately, as battery sizing requirements vary. Although it initially seems expensive with a suggested retail just under $7,000, if you need close to 5,000 watts of grid-tie power with battery backup, this is your lowest-cost option, and it certainly takes the least space. The M5 was extensively beta-tested, and has come up clean and trouble-free. Wiring access is easy and roomy for quick installations. There's no stacking ability for multiple inverters or 240-volt output (not that you should be running any 240-volt appliances on backup power anyway). Dual M5 units can be installed, and during a utility outage each will deliver an independent 120VAC source. They just won't deliver a 240VAC source between them.

Standard monitoring is very minimal...three LEDs that show if anything is abnormal, which may suit non-technical types. If you want more, there's a very good PC-based software package with simple plug-in connections that is standard with the new Plus model. The Plus model also includes a built-in battery charger and provisions for a back-up generator that were not available on first-generation models. The Beacon M5 has a CEC efficiency of 90%.

Outback Power Systems   *www.outbackpower.com*

Outback's engineering staff has a great wealth of residential inverter experience. When Xantrex bought Trace Engineering several years ago, much of the original Trace engineering staff left for greener fields and formed Outback. They now produce four U.S.-made grid-tie models with 2,500 to 3,600 watts continuous output. All models can be stacked for 120/240VAC output at double the wattage. At this time the Outback grid-tie software only allows stacks of two inverters maximum. We can install multiple stacks of two, but each stack has to output to its own separate breaker panel. This only becomes an issue with very large systems and households.

Because this is an engineer-owned company, there are a great variety of options, adjustments, and configurations. (*Leave No Possibility Behind* must be the engineer's creed.) Although all the individual bits and pieces can be purchased separately and assembled on-site, most Outback systems are delivered as pre-assembled and pre-wired Power Centers with a UL label on the assembly that makes electrical inspectors happy. This greatly speeds installation, eases inspection, and delivers compact packages that are more aesthetically pleasing. Power Centers weigh 145 pounds (66 kg) with one inverter, or 215 pounds (98 kg) with two inverters.

Outback inverters have quickly become the standard for battery-based inverters with a clean, strong waveform, a solid, trouble-free design, and great customer support. Their grid-tied Power Center packages include the awesome

Outback's FlexWare Power Centers are available with one to four inverters, more options than you can imagine or will ever need, and can be purchased pre-assembled or as individual parts.
PHOTO: OUTBACK POWER SYSTEMS

Outback MPPT charge control in either a 60-amp or 80-amp version, with monitoring software for grid-tie that tracks instant, daily, and cumulative power generation.

Outback inverters with the optional grid-tie software will automatically sense when utility power is available and go into grid-tie mode to sell any incoming DC power beyond what's needed to trickle the batteries at a float charge. If the utility fails, they'll switch to battery power in about 20 milliseconds under most conditions. This is so fast you probably won't notice it happening. Grid-tie Power Centers include the Mate, a remote monitor and control panel that also has an RS-232 output for computer monitoring. CEC efficiency is 91%. Standard warranty for Outback grid-tie components is 5 years.

Author Honesty, a Disclaimer: I've known and worked with Outback products and people for almost two decades. They're open, honest, and supportive, but haven't forgotten to have fun along the way. I like to see them succeed, because they build good equipment, and back it up with real expertise. I participated in beta testing of the MX60 charge controller, and the grid-tie inverters; both outstanding pieces of hardware/software. Outback inverter serial numbers GFX0001 and GFX0002 live on my shop wall and have been doing flawless duty making my PG&E meter spin backwards.

SMA Sunny Island   www.sma-america.com

The Sunny Island is a unique battery-based inverter package that can operate on- or off-grid. The 2nd-generation North American model is rated at 5,000 watts. When the utility fails, the German-built Sunny Island uses stored battery energy to create an instant, stable, 120VAC signal within your home (or neighborhood, potentially) that allows any conventional direct-tie inverters to keep operating. So rather than connecting your PV array to a charge controller for battery charging, the Sunny Island connects the PV to a direct grid-tie inverter like the Sunny Boy. If the utility power fails, the Sunny Island uses battery energy to create

Sunny Island inverter for on- and off-grid needs.
PHOTO: SMA SOLAR TECHNOLOGY AG

grid-like power that's obviously going to keep your appliances happy, and fools any direct-tie inverters into thinking the grid is still up. The direct-tie power will be used to run your house, and if there's still power available, recharge the Sunny Island's batteries. Once the batteries have recharged, and if there's still excess power available (and the utility grid hasn't come back), the Sunny Island will shift the 60-hertz frequency slightly to turn off the direct-tie inverter(s). If it's nighttime, or there's insufficient PV power being input, then stored battery power makes up the shortfall. If batteries get low, an onboard generator management system will automatically start your backup generator. If the generator fails, or there isn't one, a load-shedding relay can turn off selected non-critical loads when the batteries reach a slightly lower point.

The Sunny Island is particularly good if your PV array needs to be a long distance away, since all power transmission is at 120VAC or 240VAC. It can also be used to create a grid, where none exists, for several homes or outbuildings in a remote neighborhood. Multiple

Long Distances?

If your PV array must be located a long distance from your home, or if you want to create your own grid for outbuildings, the Sunny Island is ideal.

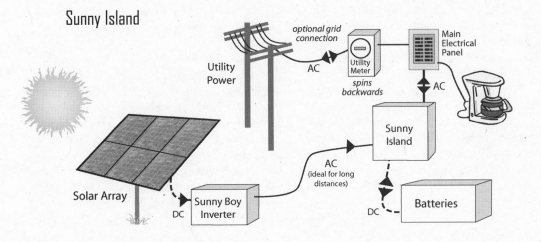

Sunny Island

direct-tie inverters and multiple Sunny Islands can all be tied together in a large grid covering many square miles. This could be just the ticket for a big, spread out eco-resort.

The Sunny Island delivers its listed wattage continuously, or two to three times that wattage for progressively shorter periods. Two units can be stacked for 240-volt output, and multiple single or stacked units can coexist on a single grid system. Because the Sunny Island doesn't work directly with a PV array, it doesn't qualify for rebates in most states. (But the Sunny Boy connected to the PV array does qualify.)

The Sunny Island is the easiest way to add battery backup to an existing or planned grid-tie system. It will work fine with other direct grid-tie inverter brands like Fronius or PV Powered. It's also the smallest and least intimidating battery-based inverter package, but it is the most expensive.

Lightning Protection

Some sort of lightning protection is a good idea on any of these systems. How seriously you consider lightning protection depends mostly on where you live. Californians can take a pretty relaxed attitude; Floridians need all the help they can muster.

Lightning wants to go to the earth. Your job is to give it an easy, obvious path, and then stay out of the way. In all cases, lightning protection begins with grounding the nice electrically transmissive aluminum frames of your modules; not the mounting structure—you want to ground right to the module frame. This ground needs to connect to the main household ground rod. If there's a secondary ground rod at the array, that's fine, but all ground rods in the system need to connect together. Good grounding will divert better than 90% of strikes harmlessly.

Beyond grounding there are inexpensive $50 absorbers (also called lighting arrestors; see photos on the next page). Delta is a popular brand; PolyPhaser offers better protection for those solar systems in serious need of lightning protection. Your installer should have a good idea how much of a threat lightning will be to your system. Check out this website to see a great map of lightning density in the U.S. www.lightningsafety.noaa.gov/lightning_map.htm

Safety Equipment

Every renewable energy installation is going to have a few bits of system-specific safety equipment in the form of combiner boxes, disconnects, fuses, ground-fault protection, and a few other gizmos that quickly cause your eyes to glaze over and start skipping ahead. There's no sex appeal in safety, so we'll cover this quickly. Just remember...this stuff keeps your house from burning down! (Got your attention now?)

DC Combiner Boxes PV modules are wired in "series-strings." A string might only be one module for a 12-volt battery system, or it could be up to 24 modules for some direct grid-tied inverters. Generally, if there's two or less series strings in a system, an unfused combiner box is used to wire the strings together. With three or more strings, a fused combiner is used, since if one string or module is shorted out, there's the potential for more amperage from the combined strings than the module wiring or J-box is rated to handle. Combiners may be roof, rack, or wall-mounted.

ABOVE: An Outback combiner box with a lightning arrestor on the left side. *PHOTO: DOUG PRATT*

BELOW: HU361RB disconnect (right) and a Sunny Boy inverter with small lightning arrestor on the bottom (left). *PHOTO: SUMMIT ELECTRICAL*

DC Disconnect Electrical code requires a manual disconnect between any power source and any appliance. This is so the appliance can be removed for service or repair without leaving hot wiring exposed. In RE systems, power sources include the PV array, wind turbine and battery bank. Appliances are the inverter and, in battery systems, the charge controller.

For battery-based systems with lower voltages, the disconnect is simply a couple of $40 circuit breakers on either side of the charge controller, and a bigger more expensive circuit breaker between the battery and inverter.

We Don't Need No
Stinkin' Disconnect!
In November 2006,
PG&E officially stopped
requiring an AC
disconnect on any
grid-tie system with a
conventional electric
meter. Reason prevails!
(They'll just pull your
meter if they're feeling
particularly paranoid.)

For direct grid-tied systems we're usually working with much higher voltages, typically 250 to 550 volts, so more specialized equipment rated for up to 600 volts is called for. Aesthetic it isn't. Safe and necessary it is. The Square D brand HU361RB disconnect has a special UL listing for high voltage DC. It's that gray box with the handle on the side next to many of the grid-tied inverter pictures we've shown in this book. With single-channel inverters up to about 3,000 watts, a single HU361RB will service up to three inverters.

Sunny Boy and Solectria inverters of 3,000 watts and larger now include a fused DC-combiner for up to four strings, plus a DC disconnect switch. The Fronius IG Plus series has a built-in combiner for up to six strings, plus a DC disconnect switch. These all-in-one inverters are slightly more expensive, but are highly favored by installers and homeowners for faster installations and better looks.

AC Disconnect It isn't required by electrical code, or UL listings, but some utility companies require a lockable AC disconnect between their meter and your grid-tie device. This is so they can positively lock your grid-tie off if they're doing repairs in the area. It's a little baffling and frustrating as to why, since UL1741 (which every grid-tie inverter is certified to), has several layers of shut-off protection in case of utility failure. But it's their grid, and you gotta play by their rules. For most residential systems this requirement will cost you under $100. If you have multiple inverters, you may have to install a small AC circuit breaker panel as an AC combiner box, since the AC disconnect is required to be a single lever.

The other sticking point here is that the disconnect usually has to be within 10 feet of the utility meter. Occasionally we've run into situations where the meter is down by the county road, while the house is several hundred yards up the hill. Rather than trench to the meter and run power down and back, the practical solution is to simply install a larger disconnect that will shut off the entire service to the house. The chance of the utility ever actually using the AC disconnect is infinitesimally small.

That concludes our safety section; back to our regular programming. Thank you for staying awake. You're officially safe now.

Wind Turbines

Big, heavy, and slow may not sound like the ideal partner, but it's what we like to see in wind turbines. A big blade will catch more energy during the mild winds that most of the U.S. experiences, thus producing more power every day. Since it's the single-most important specification, we've listed the blade sweep area in square feet for each turbine. Nothing else will tell you as much about a turbine. Heavy construction means the machine won't be so stressed during storms, and historically, the heavier the wind machine, the longer its life expectancy. Slow refers to the rotor speed. Slower machines are much quieter, bearings last longer, and there's less stress through the whole drivetrain. Now, the reality is that all these desirable features drive the price way up, so we need to reach a compromise between the ideal, and what's affordable. Let's look at what's typically available for residential use (*listed alphabetically*).

Abundant Renewable Energy   *www.abundantre.com*
I'll admit it upfront—this Oregon manufacturer comes closest to my ideal of big, heavy, and slow. They currently offer the ARE 110 model that has a sweep area of 110 square feet (10.1 m^2), which is large for its rating of about 2,500 watts. The tower top weight is 315 pounds (143 kg), also huge for a turbine of this power. They've only been available in the U.S. for a few years, and they've been upgrading bits and pieces as they've gone

Blow Pollution Away

Over its approximate 20-year life, a small residential wind turbine can offset approximately 1.2 tons of air pollutants and 200 tons of greenhouse gases.

ARE 110 Model.
PHOTO: ABUNDANT RENEWABLE ENERGY

Wind Kilowatts!

During 2008 in the U.S., commercial wind turbines generated about 49 billion kilowatt-hours (166 trillion Btus), enough to serve more than 5.7 million average U.S. households. The industry growth goal is 100,000 megawatts installed by 2020.

SOURCE: www.awea.org

along, fine-tuning the design. Everyone who has flown one has good things to say about them, particularly that they're VERY quiet and dependable. The tower kits offered by the North American distributor come complete with anchors and even a hoisting cable. ARE turbines are available for battery charging applications with a charge controller and diversion load, and also for direct-tie systems utilizing a Windy Boy inverter. Cost-wise, ARE machines are near the upper end of the spectrum. Standard warranty for ARE turbines is 5 years.

Bergey Windpower   *www.bergey.com*

Bergey has been building heavy, dependable wind turbines since 1980. They pioneered the simple side-furling design that practically all turbines use for overspeed protection now.

Bergey offers their small XL.1 with a 52.8 square feet (4.9 m²) sweep area and a maximum output of 1,300 watts in 24-volt DC only. (A 48-volt unit and a direct grid-tie unit are under development.) This 75-pound (34 kg) turbine was engineered in Oklahoma, and is manufactured in China. It's a good unit, but don't expect a lot from the modest blades unless you've got average winds over 10 mph. Their tower kits are some of the best in the business. They offer tilt-up guyed-tower kits from 60- to 104-feet (18.3- to 31.7-meters).

Bergey also has their larger Excel turbine with a 380-square foot (35.3 m²) sweep area and a maximum output of 10,000 watts in the grid-tie model, or 7,500 watts in 48-volt DC (*shown on page 71*). This turbine has been in production since 1983, with hundreds of units installed. Built in Oklahoma, it's heavy at about 1,000 pounds (454 kg) and dependable. If you've got serious wind, this is a serious turbine with any bugs long since worked out. The Excel comes with a controller or grid-tie inverter, depending on model. Guyed lattice tower kits are available from 60- to 140-feet.

Bergey machines are affordably priced, but that doesn't mean they're shoddily built. Anything but. I've had an XL.1 churning away for years atop my 53-foot tower, enduring several gusts over 100 mph (160 kph). So far I've been nothing but impressed. Standard warranty for Bergey turbines is 5 years.

Kestrel *www.kestrelwind.co.za*

Kestrel wind turbines are the new kids on the block, and by all indications they're here to stay. Begun as a small South African enterprise in the 1990s, the company was later bought out by Eveready (yes, the battery people) and brought to a whole new level of sophistication. With considerable resources now at their disposal, Kestrel built their own wind tunnel to test and perfect their machines. (Why rely something as fickle as Nature if you don't have to?)

Four models are currently available for residential wind farmers. Like the turbines built by Proven Energy *(see page 119)*, all Kestrel turbines utilize axial-flux alternators to harvest the wind's raw power. Unlike the standard radial-flux alternators found in most commercially available machines (i.e. Bergey, Southwest Windpower, ARE, etc.), in which a drum containing fixed magnets rotates around a cylindrically shaped stator, axial-flux machines utilize large disks (in which the magnets and windings are embedded) to produce their power. The result is a design with less internal magnetic drag, which allows them to produce more wattage in the light to moderate winds most people experience most of the time.

The two smaller models, the 600-watt e150 and 800-watt e220, are both battery-charging machines available in 12, 24, 36, 48, 110, and 200 volts DC. The e150 has six blades designed to turn briskly in moderate winds, but interfere with each other in high winds. This interference-created turbulence serves as braking mechanism. The three larger models are all three-blade designs with passive mechanisms to change the blade pitch (turn the airfoils away from the wind) as the turbine speed increases.

The Kestrel e400[i] (3,000-watt) turbine's compound blades offer good low speed and high speed performance.
PHOTO: KESTREL

For direct-tie applications, the e300[i] (1,000 rated watts) and the e400[i] (3,000 watts) are both designed to work with Windy Boy inverters when purchased in higher voltage ranges, although lower-voltage models for battery charging are also available. The e300[i] weighs in at 165 pounds (75 kg) with a 76-square-foot (7.06 m2) sweep area. Its big brother, the e400[i], is a hefty 484-pound (220-kg) turbine boasting a sweep area of 135 square feet (12.56 m2). Rated-watt per rated-watt, Kestrel machines are a little pricier than the popular Bergey and Southwest Windpower turbines, but their enhanced performance in light to moderate winds usually makes the Kestrel the wiser choice. Kestrel offers 2-year warranties on their turbines.

Kestrel also offers a couple of novel guyless tower designs, including a tripod pipe tower (it looks just like it sounds) and a

Comparison of Popular Wind Turbines

	Kestrel e150	Bergey Windpower XL.1	SouthWest Windpower Whisper 200	Kestrel e300[i]	Southwest Windpower Skystream
Rated Power	600 watts	1.0 kW	1.0 kW	1.0 kW	2.4 kW
Cut-in wind speed	6.25 mph	5.6 mph	7.0 mph	6.7 mph	8 mph
Rated wind speed	29 mph	24.6 mph	26 mph	21 mph	29 mph
RPM @ rated output	800 rpm	490 rpm	900 rpm	650 rpm	330 rpm
Approx. monthly kWhs @ 12 mph	75 kWh	188 kWh	158 kWh	200 kWh	350 kWh
Rotor Diameter	4.9 feet	8.2 feet	9.0 feet	9.8 feet	12.0 feet
Maximum design wind speed	120 mph	120 mph	120 mph	120 mph	140 mph
Turbine Weight	66 lb.	75 lb.	65 lb.	165 lb.	170 lb.
Direct Grid-Tie	no	no	yes	yes	yes

METRIC CONVERSIONS: mph x 1.6 = kilometers / hour | feet x .3048 = meters | pounds x 2.2 = kilograms
SOURCES: Bergey Windpower, Southwest Windpower, Solar Wind Works, Abundant Renewable Energy, Home Power and DC Power Systems

scissor tower that functions like a teeter-totter leveraged on a base standing just a little higher than either of the two arms. Think of the turbine as the fat kid on one end of the teeter-totter. To win the game (and make the system operational) you have raise him to the vertical position. Just like you always dreamed of doing.

Proven Wind Turbines   *www.provenenergy.com*

Proven turbines are built in Scotland, where they have an intimate familiarity with some of the world's most brutal wind conditions. For residential use, these are the toughest, most durable turbines money can buy. And yes, you're going to pay for that durability. Using all non-corrosive, marine-quality parts, these turbines are the ultimate in big, heavy, and slow. Having no tail, they are all

Proven WT2500 wind turbine. *PHOTO: PROVEN ENERGY*

Comparison of Popular Wind Turbines (continued)

	ARE 110	Proven 2.5	Kestrel e400[i]	Bergey Excel
Rated Power	2.5 kW	2.5 kW	3.0 kW	7.5 kW (DC) 10 kW (AC)
Cut-in wind speed	6.0 mph	6.0 mph	6.25 mph	7.0 mph
Rated wind speed	25 mph	26 mph	24.6 mph	31 mph
RPM @ rated output	250 rpm	300 rpm	500 rpm	300 rpm
Approx. monthly kWhs@ 12 mph	410 kWh	415 kWh	340 kWh	900 kWh (DC) 1,090 kWh (AC)
Rotor Diameter	11.8 feet	11.1 feet	13.1 feet	23.0 feet
Maximum design wind speed	100 mph	145 mph	120 mph	120 mph
Turbine Weight	315 lb.	418 lb.	484 lb.	1,000 lb.
Direct Grid-Tie	yes	yes	yes	yes

METRIC CONVERSIONS:
mph x 1.6 = kilometers / hour | feet x .3048 = meters | pounds x 2.2 = kilograms

How High?

California allows towers of at least 65 feet on any property of one acre or more, and at least 80 feet on 5 acres or larger. Many states still limit tower heights to 35 feet, which severely limits power production.

downwind machines, with the blades downwind of the tower. The blades have a unique polypropylene hinge built into them that allows the blade to cone inward as wind speed increases. So the higher the wind speed, the smaller the sweep area. This clever control allows continued operation under the harshest conditions. There is no wind speed at which these turbines will stop producing, up to 145 mph!

Two models are offered by the U.S. distributor for residential applications: a 418-pound (190 kg), 2,500-watt unit with a sweep area of 96.7 square feet (8.9 m^2); and an 1,100-pound (500 kg), 6,000-watt model with 254 square feet (23.6 m^2) of sweep area. Both models are for battery charging or direct grid-tie with a Windy Boy inverter. Standard warranty is 5 years. Prices fluctuate with exchange rates, but you can expect to pay a premium for these state-of-the-art machines. Proven has some aesthetically gorgeous, self-supporting monopole towers and conventional tilt-up guyed pipe towers.

Southwest's Air series wind turbine. *PHOTO: SOUTHWEST WINDPOWER*

Southwest Windpower Turbines   *www.windenergy.com*

Southwest Windpower, which has been making small wind generators in Arizona since the late 1980s, has several models available. Their Whisper turbines tend to be small, lightweight, fast, and affordable, but if you have serious winds, you may want to consider other options.

Southwest's cast-in-one-piece Air-series turbine has been a popular unit, but with only 11.5 square feet (1.06 m^2) of sweep area, don't expect a lot of power from it. Current models are rated at 400 watts peak and have a 3-year warranty. Their powder-coated marine model is well-received, and you'll see them on sailboats worldwide. This is a high-speed turbine so it gets noisy at higher wind speeds. Tilt-up guyed pipe tower kits of 27 or 45 feet (8.3 or 13.7 m) are available; you supply the 1.5-inch (3.8-cm) pipe.

The 900-watt Whisper 100 turbine *(see page 67)* is specifically made for stronger winds. It has a relatively small blade, 38.5 square feet (3.6 m²) of sweep area, attached to a fairly large generator, and it weighs 47 pounds (21 kg). If you don't have 12 mph (19 kph) or higher average winds, this isn't your turbine. A better choice for most folks with moderate winds is the Whisper 200. It has a relatively good-sized blade of 78.5 square-foot (7.3 m²) sweep area on a modest 1,000-watt-maximum generator, with a weight of 65 pounds (30 kg). This turbine will perform well under the modest winds that most of North America sees most of the time. A grid-tie version of the Whisper 200 with a Windy Boy 3000 inverter is also available. Both the Whisper 100 and 200 models have 5-year warranties. Tilt-up guyed pipe tower kits for either model are available from 24 to 80 feet (7.3 to 24.4 m); you supply the 2.5-inch (6.35 cm) pipe.

The Whisper 500, a 3,000-watt model with side-furling, is the last of the two-bladed turbines. If your winds blow steadily from one direction, two-blade turbines are fine, but when winds are shifting, these turbines suffer a lot of heavy vibration. The 500 model weighs about 155 pounds (70 kg) and has a 5-year warranty. Tilt-up guyed pipe tower kits (hardware and cabling) are available from 30 to 70 feet (9.1 to 21.3 m). This turbine needs a 5-inch (12.7-cm) pipe for its tower, which is sometimes a difficult size to find.

Southwest Windpower's 2.4-kilowatt Skystream 3.7 is a radical departure from the company's other popular designs. For starters, it's a downwind machine, meaning that the turbine lacks a tail and therefore turns away from the wind rather than into it. It's also the only Southwest machine designed specifically for direct-tie applications (although it can be adapted for battery charging with the addition of a separate controller). But what really makes the Skystream 3.7 unique is the fact that along with the alternator, they've managed to pack a direct-tie inverter inside the nacelle. The inverter takes the turbine's inherently erratic current and outputs a grid-compatible alternating current. This does away with the need for a separate inverter on the side of the house (where, I should add, many RE professionals think

Tower Investment
Be prepared to spend at least as much for your tower system as you spend for your wind turbine.

The Skystream 3.7® wind turbine. *PHOTO: SOUTHWEST WINDPOWER*

the inverter really ought to be). The Skystream 3.7 weighs in at 170 pounds (77 kg) and its three scimitar-like blades sweep an area of 113 square feet (10.5 m²). This model also has a 5-year warranty. To hold the machine up in the sky in a fashionable repose, Southwest offers a trio of galvanized-steel monopole towers, ranging from a 33.5-foot (10.2-m) model to a 60-foot (18.3-m), 1,500-pound (680-kg) affair that tapers from around 14 inches (36 cm) at the base to a little over 6 inches (15.5 cm) at the top.

Charge Controllers (battery-based systems only)

Until a few years ago, the charge controller's job in a battery-based grid-tie system was pretty minimal. It was only there in case the utility failed on a sunny day, and the house wasn't using all the available incoming power. Only then did it get to be anything more than a simple connection that turned on at dawn and off at dusk. All the rest of the time the inverter monitored the battery voltage, and sucked off any power beyond what was needed to maintain a gentle float voltage. Then Maximum Power Point Tracking for charge controllers was developed. By running the PV array at whatever voltage delivers the greatest wattage, and then downconverting that higher voltage into amps the battery can digest, our PV array can average about 15% more energy output over the year. And since higher voltage transmits easier, this means smaller wire sizes and less transmission loss.

Given the large size of most grid-tied systems, the addition of an MPPT charge controller is like having an extra two or three modules in the system. This is an expense that more than pays for itself immediately. Outback's masterful FM60 and Xantrex's MPPT60 controllers are two of the manufacturers using MPPT technology.

Outback's FLEXmax controller with MPPT controls. *PHOTO: OUTBACK POWER SYSTEMS*

Xantrex MPPT60 charge controller. *PHOTO: XANTREX*

MPPT Technology: Squeezing the Last Watt from Your Array

Solar modules are designed to operate at higher voltages than they are nominally rated for. There are two main reasons for this. Since voltage always flows from a higher potential to a lower one, the modules in battery-based systems need to operate at a high enough voltage to charge the batteries in both low light conditions (when the array voltage drops), and when the batteries reach a high state of charge. Since lead-acid batteries (in a nominal 12-volt system) often reach potentials as high as 15.5 volts during equalization, the array's rated voltage must be even higher. Typically, this will be in the range of 16.5 to 18 volts. During the bulk charging stage, a typical charge controller will simply hook the array directly to the batteries. The batteries, of course, don't have any use for all that extra voltage, so they pull the array voltage down to a comfortable level.

MPPT (Maximum Power Point Tracking) charge controllers get around this problem by using a high-frequency DC to DC converter to provide a voltage the batteries are happy with. In the process, it uses the extra voltage to produce usable amperage in excess of what the modules were designed to produce.

In battery-based systems, power point tracking works best when there is a large disparity between the module voltage and the battery voltage. This most often occurs when the batteries are partly discharged, under a considerable load, or when the PV modules are cold.

Direct grid-tied inverters also use MPPT technology to level out the voltage changes brought on by time-of-day and weather-related changes, thus providing the inverter with a consistent voltage input. Just like in a battery-based system, MPPT squeezes the last available watt out of the array, while maximizing inverter efficiency at the same time.

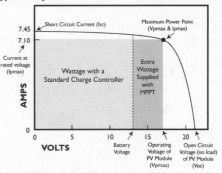

Typical Amps-Volts Curve for a 120-watt PV module

Batteries

Batteries are usually the most misunderstood system component. And, as is usual when we humans don't fully understand something, we wrap it up in fables, superstitions, and old wives tales. Let's see what we can do for enlightenment...

What's a SMALL Battery?

In the context we're talking here – battery banks that can run a house for hours – a **small** battery is anything near the size you might find under the hood of a car. In battery nomenclature, these would be Group 24, 27, or 31 batteries with up to about 100 amp-hours capacity, and weighing as much as 75 pounds each. **Medium**-size batteries go from 100 amp-hours up to around 400. Anything with 500 or more amp-hours is officially **big**.

Lead-Acid Batteries   Within the general family of lead-acid batteries there's significant fine-tuning, tweaking of chemistries, plate design, cell connections, and sizing to produce batteries that perform better under specific operating conditions. For instance, an automotive starting battery has to deliver a few hundred amps for a few seconds, then the alternator quickly recharges it. A solar battery in a remote site will be asked to deliver a steady trickle of amps for the lights, with the occasional surge for the water pump or microwave, and it might have to do this for several days without recharging. An automotive battery would suffer a short, ugly life if asked to do this, but a true deep-cycle battery will thrive for years and years. They're both lead-acid batteries, but they're built differently.

Wet-Cell vs. Sealed Batteries   Off-grid systems use deep-cycle, wet-cell batteries, and you will routinely remove their caps to add distilled water. Wet-cell batteries are designed to be charged and discharged regularly. The sun shines and we charge the batteries. The sun goes down, we turn on lights or other appliances, and discharge the batteries. Wet-cell batteries are happy doing daily discharges of 20%–30% of their capacity, but they should not be routinely discharged below 50%.

Grid-tied systems don't cycle the batteries regularly since months or even years may go by without a power failure. Sealed batteries have their chemistry tweaked to better tolerate these long periods of inactivity without losing the ability to quickly respond when they're really needed.

Sealed batteries don't have any way to put lost water back into the battery, so charging has to be very carefully controlled. Excess charging energy breaks the water into hydrogen and oxygen gases which can escape through one-way vents. And once gone, the water cannot be replaced.

Sealed lead-acid batteries have two production technologies, AGM (Absorbed Glass Mat) or Gel. The AGM type uses a fiberglass-like material with a liquid electrolyte. They're easier to produce but have less liquid and so are less tolerant of over-charging. Because the electrolyte is liquid and can move around a bit, they tend to tolerate high charge and discharge rates quite well. What they won't tolerate is high voltage: 2.35 volts per cell is the max. Higher voltages cause gassing, which vents water vapor...you get the picture.

MK's 8G31 12-volt sealed gel-type battery is a tried-and-true favorite. *PHOTO: MK BATTERIES*

Gel-type sealed batteries use a jellied electrolyte that's a bit more tolerant of occasional abuse. They're more difficult to build because no air voids can be allowed when filling the battery with gel—voids create a dead space on the plate forever. Gel batteries start their life with more moisture, so they're more tolerant of the occasional overcharge. Still, 2.35 volts per cell is usually the recommended maximum charge voltage. Although they're more expensive, gel cells usually deliver slightly better life expectancy than AGM cells.

Hawker Envirolink gel batteries have a long-life expectancy and require a forklift to move. *PHOTO: HAWKER BATTERIES*

In all cases, bigger battery cells last longer. Your battery bank will need a certain amp-hours capacity in order to deliver the backup power you need. You could build that bank out of many smaller batteries, or a few larger batteries. The bank with a few large batteries will last longer, be less prone to charging and performance problems, and cost you less per year. Count on it!

The bottom line here is that if you're adding batteries to a grid-tie system, it's important to use sealed lead-acid batteries. They may cost more initially, but will live many times longer in emergency backup service than a comparable wet-cell battery. ❖

Got Lightning? Go Solar

Neither rain, nor lightning, nor hurricane-force winds shall stay this faithful accountant from the swift completion of his appointed duties. No, John Kerr never actually said that, but it seems to be the Florida tax accountant's operating principle, nonetheless. Not only did his home survive direct hits from Hurricanes Jeanne and Frances in the fall of 2004, his home office never lost power, even though the grid was down for days at a time following each storm.

It wasn't hurricanes, however, as much as the frequent, grid-busting lightning storms around Port St. Lucie that provided the original prodding for John to install his simple-yet-functional 12-volt PV system. Consisting of six 90-watt roof-mounted Matrix modules, a quartet of Best Power 12-volt, 135-amp-hour sealed UPS batteries, and a 1,500-watt Coleman Powermate modified-sinewave

inverter, this small but adequate system powers John's accounting office and the home's security system, as well as the garage and all of the power tools John uses when he's not chasing down loopholes in the tax code. A separate 24-volt system runs the pumps that circulate hot water from the single roof-mounted solar hot-water panel. And, to keep watt-gobbling artificial lighting to a minimum, a trio of tube skylights brings an abundance of natural light into the office and the home he shares with his wife Kathleen and son, Ian.

The system is not configured to allow John's homegrown wattage to pass from his array into the Florida Power & Light grid, nor could it be, without an approved inverter. But that doesn't prevent John from drawing power from the grid—with the aid of a manual bypass switch—on those rare occasions when he needs a little extra power.

All told, it's a clever, inexpensive solution to a persistent problem. But, hey—isn't that what we expect from our tax accountants?

Top closeup photo shows the PV panels on Kerr's roof.
PHOTOS: JOHN KERR

CHAPTER TWELVE

Who Does the Solar / Wind Electric Installation?

A re you thinking more seriously about a solar and/or wind energy system yet? Good. It is something that's going to return security and comfort for many years to come. Now comes the decision of whether to hire a contractor, or do some or all of the installation yourself.

Finding and Qualifying a Contractor

If you live in a state like California or Colorado that has an active, well-funded rebate program, then finding a contractor is easy...maybe too easy. Don't be fooled by companies that have a lot of money to spend on newspaper ads and slick literature. Often the low-profile installer with low overhead can do the job more to your satisfaction, so get more than one quote and study each quote carefully. One advantage of doing a lot of solar installs, of course, is that crews not only get faster, they get better. (Oh gee, we won't make *that* mistake again!) Look in your Yellow Pages under "SOLAR" or online at *www.FindSolar.com*. You can also call your state energy office for referrals (*see listings in appendix*). Another source of quality installers is the North American Board of Certified Energy Practitioners (NABCEP). It's a mouthful, but

Do It Yourself?

If the thought of working with lethally high DC voltages up on a hot roof sounds fun, you might enjoy being a professional solar installer. The industry needs you. But if this is your own system, it's not a project for beginners. Hire a pro and be his helper.

it is the only organization currently testing and certifying solar electric installers. It is a voluntary program that is not inexpensive nor easy. So these are folks who are serious about being professional and knowledgeable. Find them, and a list of certified installers, sorted by state at: *www.nabcep.org*.

In order to give a realistic quote, a contractor absolutely needs to perform a site inspection. All houses are different, with roofs that face in every direction, covered with a variety of materials, and shaded by who knows what. If a contractor is willing to give a firm quote without at least a drive-by, you should be very suspicious. If you have any trees, buildings, dormers, or other potential obstructions you should absolutely expect someone to show up with a ladder and a Solar Pathfinder (or similar) device that will show what's going to create shading problems, and when. (Obviously the ladder is only needed with roof-mounted arrays.) The venerable Pathfinder has significant competition from several computer-based solar access gizmos, and they are rapidly becoming the assessment tool of choice for many solar companies for ease of use, speed, and accuracy.

The Solmetric SunEye uses a 160° camera lens coupled to a small PDA computer that figures out seasonal sunpaths and what is shade or sun within those paths.
PHOTO: DC POWER SYSTEMS

Payment schemes vary. Because the state agencies that approve rebates are often backlogged weeks or even months, expecting full payment at the beginning of the process isn't realistic. Someone other than you gets to sit on your cash all that time. You should expect to put down a substantial down payment of $1,000 to $5,000 and sign a contract, with the balance due on completion of the installation. If you find a contractor who's willing to accept the state rebate in lieu of payment from you, that's a real plus. Those are dollars that never need to leave your pocket. And since the contractor can't claim the rebate until the job is signed off by the local electrical inspector, you're assured a speedy, proper installation.

And finally, make sure your contractor is licensed. All but a couple very rural states require electrical workers to be licensed. You can usually find their state license number imprinted on their business card, stationary, and email. Most state contractor licensing boards have convenient websites where you can quickly check to see if a contractor's license is valid and if there are any pending complaints or actions. For a website with links to all state contractor licensing boards, visit the Contractor's License Reference Site at: *www.contractors-license.org*.

Do the Installation Yourself?

Unless you've done this before, we're going to suggest that installing a grid-tie system isn't a good project to undertake entirely by yourself. Because of the slow, complex, and sometimes baffling paperwork required, and because it requires working with lethally high voltages, we recommend that your best option for cutting the cost is to hire a pro, and become his or her helper. And even if you are a pro already, many programs have a self-install penalty. In California and New Jersey for instance, there's a 15% reduction in rebate for self-installing. So it's best to hire a contractor. With the extra 15% rebate you can afford it.

You can still shop for, and buy, all the major components yourself, saving the contractor markup on parts. There are a number of retailers willing to sell directly to end users, with varying levels of customer support. Some offer very complete stan-

Uni-Solar's peel-and-stick PV material being applied to new standing-seam metal roofing. This was a class to train dealers; installation only requires a team of three people normally. *PHOTOS: DOUG PRATT*

dardized packages, while others will customize a package to suit your available space and budget. But watch out! If you web browse for the absolutely lowest price, you can expect absolutely no service or support to go with it. You'd best know what you're doing. Also, watch out for long delivery times. The real lowball dealers won't even order your hardware until they get your check. (How do you suppose they pay for

Hands-On Guides

If you feel this is
something you can
tackle successfully, here
are a few resources to
guide you along (see
page 180 for details):

A Guide to
Photovoltaic (PV)
System Design and
Installation by Bill
Brooks of Endecon
Engineering

Photovoltaics: Design
and Installation
Manual by Solar Energy
International

Photovoltaic Systems
by Jim Dunlop

it?) With modules in short supply and everybody on an allocation waiting list, this sets you up for a several-month wait with your money already spent. Ideally, you will want a retailer who's supplying a schematic drawing of your system, offers technical support by phone and/or email, and can deliver your hardware within 3 to 4 weeks. The schematic drawing is particularly important. Not only is this your installation road map showing how everything connects together and with what gauge wire, most building departments now require one before they'll issue a permit. (And you've **got** to get a permit if you want a rebate!) If your dealer doesn't supply a schematic, guess who's going to? ❖

A typical rooftop installation with Unirac rails on standoffs supporting 20 Astropower (now G.E.) 120-watt modules. Standoffs are optional on conventional asphalt shingle roofs like this, but are higher-quality and less likely to leak. Excess rail length was trimmed after all the modules were secured. This would be a typical one-day installation for an experienced crew of two or three. *PHOTOS: SUMMIT ELECTRICAL SERVICE*

GUEST COMMENTARY BY CHRISTOPHER FREITAS

The Future of Renewable Energy for Homeowners?

What will be the most common type of renewable energy system of the future—off-grid or grid-tie? My usual answer is, "Both," meaning that there will just be more of each type, depending on the circumstances. But lately I have realized the most common renewable energy (RE) system of the future might actually be neither of these.

Take computers as a parallel example. Back in the early 1980s, the big question among computer geeks was whether computers of the future would be personal or mainframes—both were available and being used in ever-growing numbers. The interesting part is that neither solution ended up dominating. Something else did—the Internet, an amalgam of both personal and mainframe computers, as well as many things we never imagined. And when you're viewing a Web site through a wireless Internet connection, for instance, it doesn't matter what kind of computer you're using.

Renewable energy systems are similar—we really can't imagine what the future will look like. The eventual widespread system will probably end up as a mixture of both off-grid and grid-tie technologies coupled with some new ideas few have imagined yet. In the future, the terms "off grid" and "grid tie" may just be relics, just as "personal" and "mainframe" are to surfing the Web.

Blast to the Past

We don't always notice the scale of the changes that occur around us during our lives. As a young child, I remember turning over the telephone to find a sticker on the bottom that said, "Property of Pacific Telephone." When I asked my father about it, he informed me that everyone leased their telephones from the phone company and paid a monthly fee for each phone used. He emphasized that you could not just take any old phone you found and plug it into the phone lines—it might damage the phone network. I also remember that when the first fax machines and personal computers were being used, the phone company required you to have special "data" phone lines installed with extra fees attached. Consumers had no

choice in who provided their phone service—there was no competition.

Today, Pacific Telephone doesn't even exist and consumers now have a choice of phone service providers. Nobody leases their telephones from their service providers anymore. You can plug a fax or computer into a wall jack without causing the phone system to crash. And your "local" service provider might be located in an entirely different part of the country. Yet it all still seems to work fine and, thanks to competition, prices have dropped. The latest trend is to forego traditional hard-wire phones at home. Instead, many people use cell phones or Internet-based phone services. Both wireless and networked solutions coexist side-by-side and also work together.

Fast Forward to the Future

Today's utility companies remind me a lot of the old phone companies. They have many people convinced that permission is needed to connect renewable energy sources to "their" grid, which was actually paid for by the consumers under the utilities' monopoly status. The utilities view renewable energy as a competitor—something they need to control if they have to allow it. And, just like the telephone companies used to do, they want to require special rules and additional fees for using RE. At some point, utilities will realize that they should not only allow the widespread use of renewable energy, but that their very exis-tence requires that they embrace it or face extinction, just like many of the old telephone companies that were slow to change.

The most common RE system 20 years from now might be something that is just as unimaginable as today's Internet was three decades ago. Future-flexible RE systems might connect to a utility network when it makes sense to but also be able to work independently. There might even be independent "open" power networks similar to today's wireless networks, where people can distribute the energy produced by their RE systems to their neighbors or provide RE access to those less economically able to afford it. And the role of the utility in the future might be very different than it is now—if they still exist at all.

Beyond utilities, getting people to embrace and utilize renewable energy will require a completely different relationship with electricity—how it's consumed and what its role is. This might sound like a quantum leap, but changes happen all the time—in fact, they are inevitable. The renewable energy industry needs to seek out, invent, and take advantage of all new opportunities to reach more people and make RE a part of everyone's daily lives.

Christopher Freitas is a cofounder of OutBack Power Systems, a U.S-based manufacturer of power electronics for RE applications (www.OutbackPower.com). He has worked in the renewable energy industry since 1985.
ARTICLE FIRST APPEARED IN HOME POWER MAGAZINE, ISSUE #128

Heating Your Home with Nature's Free Energy

Passive Solar

Solar Hot Water

Geothermal Heating/Cooling

CHAPTER THIRTEEN

Using the Sun's Warmth to Heat Your Home

According to the U. S. Department of Energy, heating and cooling account for 50% – 70% of the energy used in the average American home.

E nergy is slippery stuff, and that makes it a problem. Your house loses heat at night for the same reason an ice cream cone melts all over your hands and drips into your lap: everything wants to become the temperature of the surrounding environment. Back in the days when oil, coal, and natural gas were plentiful and cheap as dirt, and no one had ever thought to place the words 'global' and 'warming' side by side, energy's elusive tendencies were usually not much more than a mild annoyance. But as Bob Dylan reminds us, *the times they are a changin'*. What was considered adequate building insulation in the days of cheap energy is wholly unsatisfactory today. Insulate, insulate, insulate! And while we are told that sealing tightly around doors and windows and ductwork makes good sense, it really does pay off in energy savings. Designing and building a house today without carefully considering how best to make it energy efficient is like deciding you'd like to make charitable contributions to your local utility company. Every month. Forever.

Passive Solar Design Tips

For those planning to build a home, every dollar you spend on good home design will save you twice that on renewable energy equipment. So the old proverbial drawing board is the place to start if you hope to make the most of the energy you use.

Orientation and Floorplan Considerations A home stretched out along an east-west axis exposes a maximum amount of exterior wall surface area to the warming rays given off by the winter sun as it tenaciously patrols the southern sky. There is, however, a little room for deviation. Thermally speaking, the heating effect is nearly the same if the house is aligned within 15 degrees either way of that imaginary east-west line, but the closer you can get to a true east-west orientation the better. The length of the roof's eaves is also very important for internal temperature control and should be calculated according to the latitude and type of climate you'll be living in, since the eaves determine which month of the year the sun no longer shines directly inside the house.

Following this same line of logic a bit further, it only makes sense to place the most frequently occupied rooms on the south side of the house where the daytime lighting is free and conserved sunshine (which is to say, stored heat) will provide heat well into the night. Keep the utility spaces and spare bedrooms on the cold, dark north side. Your in-laws will expect that sort of treatment anyway, and it will make their visits a little shorter.

Thermal Mass & Trombe Walls Of course, to really take advantage of the sun's low angle in winter, it's not enough just to point your house's broadside in the sun's direction and hope for the best; you have to invite it inside where it can really work some magic. This is because light coming in through a home's windows is absorbed by things inside the house and reemitted as heat (which is simply light with a longer wavelength) that cannot easily pass back through the glass. This is the principle behind the much-publicized greenhouse effect. To ensure that this free heat is not squandered on furniture,

Get Audited

A home energy audit is the best way to find out where the energy in your home is going and how to use energy more efficiently. Using cool diagnostic tools like blower door tests and thermographic scans, energy auditors will find your home's weak spots while determining the efficiency of your heating and cooling systems. They will also show you how to conserve hot water and electricity. Call your local utility and ask if they do professional energy audits. If not, they should be able to recommend companies that do.

carpets and dogs (none of which is all that good at storing heat), you want to place things in the path of the light that have appreciable thermal mass. What sorts of things? Well, the most obvious is ceramic floor tile. Tile can soak up a surprising amount of heat, and if it's laid over regular or gypsum concrete, as it would be for a highly efficient hydronic in-floor heating system (hint, hint), the heat gain is further enhanced.

A Trombe wall is also a terrific heat-storing design feature. Consisting in principle of nothing more than a concrete wall, or wall segment, installed within an inch or two of a south-facing window and sealed around the edges, Trombe walls are showing up in more and more homes these days, and for good reason: it takes the heat trapped between the wall and the glass 8 to 10 hours to pass through an eight-inch thick concrete Trombe wall into the interior of a building, so you'll just start feeling the heat of the noonday sun a little before bedtime. It's the stuff sweet dreams are made of.

Other indoor heat reservoirs that are effective include rock and/or concrete planters, or even large water-filled metal columns coated with unreflective paint. You can let your imagination go wild on this one.

Windows Windows are also important, in all of their particulars, as their placement can either enhance or hinder the performance of an otherwise energy-efficient house. Generally, you want to put as few windows as possible on the north, plenty of windows on the south, and more windows on the east side than on the

Finding the Best Windows for Your Home

To determine which windows are the best match for your new home, visit the Efficient Windows Collaborative at *www.efficient windows.org*. Their user-friendly Window Selection Tool will instantly provide you with efficiency data for thirty or more different window types for any U.S. city you choose. You will quickly see the disparity between good windows and mediocre ones in terms of how much extra you will spend to heat and cool your house by choosing one type of window over another.

What you won't see is how much you are helping the planet by making the right—and ultimately, the economical—choice, but this may help: every dollar not spent on electricity saves the production of 20 pounds (44 kg) of carbon dioxide (CO_2), and every gallon of propane not burned prevents the release of another 12 pounds (26 kg). And all the while you're not producing CO_2 you'll be basking in the comfort your new windows will provide.

west side. This latter consideration relates more to summer cooling than winter heating, since the summer sun that begins the day as a welcome, warming beacon metamorphoses into a Dantean blast furnace by mid-afternoon. And windows unshaded by eaves just encourage it.

Unlike the old days when windows consisted of one or two panes of untreated glass, there are now numerous glazing options, including windows coated with various metallic oxides, with the spaces between layers filled with sluggish, inert gases such as argon or krypton. These are all good developments, even if they can be a little confusing. Fortunately, when it comes to choosing the energy-conserving performance of various types of glazing, it all boils down to just two factors you need to consider: the U-factor, which is a measure of a window's heat emissivity (low-E, for instance, equals a low U-factor, which in turn means better heat retention inside the structure); and the solar heat gain, which is the amount of solar radiation permitted to pass through a window from the outside. Thus, on the south side of a house in a cold or temperate climate you would want low-E windows with a high solar gain, while on the west side of the same house you should consider using low-E, low solar-gain windows, to block the incoming radiation during hot summer afternoons, while retaining heat through the winter night.

Natural Cooling   After spending all winter conserving heat, now comes summer and a burning obsession to rid yourself of it. To do this naturally, without energy-intensive central air conditioning and evaporative (swamp) coolers, your best bet is to design your new home to take advantage of the natural convective movement of air from a low, cool entry point to a high, warm exit point. The former could, ideally, be vents or windows on the north side of a basement, while the latter might be operable skylights in the loft ceiling, or windows or vents high on the gable ends. However you do it, the trick is to provide a clear path for the air to flow, such as adjustable vents

Window Construction

Window **frame types** are pretty straight forward: aluminum is at the bottom of the energy efficiency list because of its well-known conductive properties, while vinyl and fiberglass sit comfortably at the top. But don't despair; in terms of efficiency, wood and wood-clad windows run a commendably close second.

The number of **glazing layers** is a no-brainer: the more the better, no matter where you live. Typical top-of-the-line windows have three.

between floors. Another trick to maximizing natural convection is to draw the north-side intake air across a backyard pond, preferably one with a small, simple pump that produces a modest spray of water. Air passing through the spray will give up much of its heat to the water, making it that much cooler when it enters the house.

Solar Air Panels

As much as we love them, windows and skylights present a two-fold problem. Even the best windows allow heat to leak out of the house at night, thus offsetting to some extent their daytime heat-trapping properties. And, conversely, it's difficult to regulate how much heat they're allowing into the house when the sun is shining, even with special glazing options.

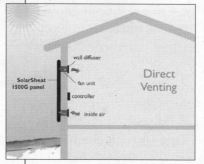

The SolarSheat 2Pak, which can be installed by the homeowner, has a built-in solar PV panel to power the fan. Two 5-inch holes are required for air circulation.
PHOTO & DRAWING: YOUR SOLAR HOME INC.

What else is there for solar space heating? Solar air-heating panels mounted on the roof or onto the south side of the house is one solution that is gaining popularity. Similar to a flat-plate solar hot-water panel in size, appearance and function, a closed-loop solar air-heating panel draws air from inside the room into the lower part of the glazed panel and, with the aid of a fan, blows warm air back into the room after it has circulated through the absorber.

Many of these units use a low-wattage fan that is powered by its own small solar-electric panel, thus doing away with the need for electrical hookups. All that's required to install a solar air collector is to poke a couple of holes through the wall or roof for the ductwork.

Then you're off and running. A wall-mounted digital thermostat tells the fan to shut off when the room is sufficiently heated.

But if complexity is your bag, don't despair. With a bit more work, individual or multiple units can be patched directly into your home's ductwork, pre-charging the furnace's cold-air-return with solar-heated air, thus reducing the amount of time the furnace has to run during each cycle.

Either way you do it, it's a great way to reap the benefits of windows without the aforementioned downsides. Of course you might think a large, black panel attached to the side of your house is something of an eyesore, and if that's the case then a roof installation might make more sense for you. Wall-mount installations are fairly simple, so long as you don't have to tap into the home's wiring or ductwork; roof mounts require a bit more labor and expertise.

There are several U.S.- and Canadian-made models of solar air panels on the market. Before buying, compare output Btus and airflow rate, ask for customer referrals in your area, and check out the warranties. An SRCC certification is a good indicator the panel you are buying is a good one.

Expect to pay on the order of 15 to 25 cents per hourly Btu. The good news is that manufacturers estimate your investment can be recouped in 4 to 8 years, depending on energy prices. With so few working parts, a solar air heating system will last a very long time. ❖

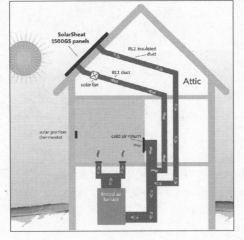

For More Information on Green Building

Florida Solar Energy Center www.fsec.ucf.edu

NAHB National Green Building Program www.NAHBgreen.org

NREL's Buildings Research www.nrel.gov/buildings

U.S. Green Building Council www.usgbc.org

Larger, multiple-panel systems which connect with the home's forced-air furnace should be professionally installed. *DRAWING: YOUR SOLAR HOME INC.*

Tropical Year-Round Greenhouse at 9,100 Feet Makes For Comfy Living

The 2,800-square-foot, two-story house built by Deborah and Mike Overmyer in the Colorado Rockies has some amazing characteristics. Constructed of glass, stone and cedar, their home's south wall is a large greenhouse (36 by 10 feet, and better than 20 feet in height) enclosing a tropical ecosystem. Oriented on an east-west axis, it remains comfortably warm, even during the frequent bouts of subzero weather common at high altitudes (9,100 feet).

Separating the greenhouse from the main house is a massive stucco-covered, concrete Trombe wall that soaks up heat from the sun during the day, and gives it back at night. On cold, cloudy days when the sunshine is reduced, the process is reversed—the woodstove and baseboard hot-water-register heaters in the great room warm the Trombe wall from the inside, allowing heat to pass into the greenhouse. But the Trombe wall is more than just the heat exchanger. As Deborah explains, "The wall has windows that open into the house and main room, giving us air circulation and natural lighting."

All of the greenhouse windows are triple-glazed, argon-filled, high-solar-gain, low-E windows designed to let in lots of light and allow very little heat to escape. Additionally, the greenhouse is covered with a

highly insulated roof, the eve of which extends out just far enough to prevent direct sun from hitting the south-side windows during the summer. And to further prevent the house and greenhouse from overheating, there is a natural convective flow of air regulated by rows of vent windows at both the floor and ceiling. With 20 feet of vertical distance to travel, the swiftly rising air creates a strong current that ensures a constant movement of air through both the greenhouse and the main house.

But even with all these carefully thought-out features, the greenhouse would still be in danger of overheating in the summer were it not for the fact the house wraps around it on the east and west sides. This prevents any morning or afternoon direct sunlight from entering the greenhouse from May to September, while allowing considerable direct light in the coldest months of the year.

All the parts add up into one very impressive whole: a greenhouse that is a net producer of useful energy. It allows the Overmyers to have a spacious, verdant, year-round tropical greenhouse in an environment most would think far too harsh and challenging for such a thing to be possible.

PHOTO: LAVONNE EWING

CHAPTER FOURTEEN

Get in Hot Water and Love It

S olar hot water systems (also called solar thermal systems) are the low-hanging fruit on the renewable energy tree for both new and existing homes. For many households solar hot water may be a better, more productive place to put your hard-earned dollars than solar electricity. Now, I can hear you saying, "But I've already got a perfectly good water heater that meets my needs. Why would I want to invest in a solar heater?" The hard economic answer is zero operating cost. The soft green answer is it's a free renewable energy source, with no hidden costs passed on to our kids. In most North American climates a modest (under $7,000) solar hot water system will supply 70%–80% of a typical family's hot water needs per year. A solar hot water system will pay for itself in less than 10 years, and if you're currently using an electric water heater, probably in under 5 years.

About 20% – 40% of a typical household's heating bill goes to heating domestic hot water (for showers, laundry, etc.).

How's It Work?

We all know that anything sitting in the sun gets warm, more so if it's a dark color. Solar hot water panels do a number of things to optimize this heat collection, with control and pumping systems that move the heated water

What to Expect

In most North American climates a modest solar hot water system (under $7,000) will supply 70%–80% of a typical family's hot water needs per year.

to storage for later use. Any solar hot water system you install would likely use your existing water heater for backup since there are obviously times when solar water heating will fall short (and we are NOT suggesting you give up showering!).

Solar pool heaters are the simplest design since they just need to add a few degrees of heat to many, many gallons of water; quantity is more important than high temperature. So pool heaters use large uncovered mats to collect the heat. This mat/collector is only a few degrees above ambient air temperature so the few Btus that radiate from the mat into the air aren't usually worth the cost of enclosing the collector. Pool heaters typically use the existing pool filter/circulation pump to divert a portion of the flow through the collector mats.

For domestic hot water, on the other hand, we want higher temperatures, but not nearly so many gallons of water heated. So domestic solar collectors will sit inside an insulated box with some kind of glazing (preferably tempered glass) to prevent heat loss. In addition, good collectors will be painted with a selective coating. This hi-tech coating is very good at accepting incoming heat, and very stingy at letting it back out.

Can I Heat My House with Solar Heated Water?

This question always comes up...why not just get a whole bunch of solar hot water collectors and use them to heat your house? Well, yes, you could do that, but in most North American climates that's probably not a smart idea. You're trying to extract a resource (solar heat) at the exact time of year when that resource is in shortest supply. So you'll need a LOT of solar collectors, which are expensive, and they're going to sit doing nothing for you seven to nine months of the year. Typically not a good investment.

On the other hand, household hot water needs are very consistent throughout the year. So your investment works for you all year long, and it's a much more modest investment to start with.

If you really are interested on going big into solar hot water, get Bob Ramlow's *Solar Water Heating: A Comprehensive Guide*. Bob is a Wisconsin native, and does some amazing and highly intelligent things with solar hot water collection systems in cold climates.

Collection Starts Here

Domestic collectors come in three basic configurations: batch, flat plate, or evacuated tube. Which type is best for you depends on your climate and what you're expecting your system to deliver.

Batch collectors are the simplest type and usually have the lowest installed cost by a wide margin. A batch heater is basically a tank of water in an insulating box with a clear cover. The tank may be painted with a selective coating, and the collector might get "flattened out" a bit by using a series of large pipes, but the idea is the same. If your water gets sun-warmed sitting in a black tank before it gets to the cold water inlet of your regular water heater, you save heating energy. No pumping or control systems are needed with batch heaters. Being open-loop systems, they're simply plumbed inline ahead of your existing water heater, which then becomes the backup heater, only running as much as needed to bring the water up to temperature. Batch solar heaters are simple and less expensive than flat plate or evacuated tube systems. But they cannot be made freeze-proof, so there's a limited market for this type of solar collector in North America.

Flat plate collectors feature a large black collection area inside

an insulated box. Typical sizes are 4-foot wide by 8- to 12-foot long (1.2-m wide by 2.4- to 3.7-m long). They have a series of small pipes with large fins on either side to collect heat. Because the pipe size is small, flat plate collectors

ABOVE: ProgressivTube® is one of the best and least expensive batch-type heaters. It consists of several 4-inch (dia.) pipes that absorb heat and store the hot water. Available in various sizes, from 18 to 50 gallons. *ILLUSTRATION: THERMAL CONVERSION TECHNOLOGY*

LEFT: Four of this home's 12 flat plate collectors are mounted above their garage. This large system provides both domestic hot water and home heating, plus enough for an outdoor hot tub. A good use for the old-style collectors that were destined to be recycled. *PHOTO: LAVONNE EWING*

rarely hold more than a gallon of fluid. Circulation, then, is critical: collect the Btus and move that heat into storage. Copper is the usual choice for the plumbing and fins, with selective-coating finishes being very important. Flat plates are a practical choice for freezing climates (more on that below), are more aesthetically pleasing, and generally do a better job of collecting heat than batch collectors. But they require pumping and control systems which increases the system cost and complexity.

Evacuated tube (ET) collectors are a newer and still-evolving technology. Each individual unit consists of two glass tubes, one inside the other, separated by a vacuum. Kind of like a long skinny thermos bottle without the metal skin on the outside. Inside this tube-within-a-tube is a copper heat pipe attached to a black copper absorber plate. The heat pipe is filled with a liquid (generally a refrigerant, such as alcohol) that is converted by sunlight into hot gas. The gas wicks to the top of the tube where it gives up its heat to a heat-conducting manifold before condensing and flowing back into the heat-gathering end of the heat pipe. Water or glycol passing through the manifold (which holds from 30 to 60 individual tubes) ferries heat away to the solar storage tank. The vacuum space makes the best possible insulator, so ET collectors really excel at hanging onto heat once they've collected it. Because of their excellent heat retention, ET collectors are favored in cooler, cloudier climates, or where higher water temperatures are desired. Generally ET systems are more expensive than their flat-plate counterparts.

A home in Laytonsville, Maryland uses 32 Sunda evacuated tubes to heat their water. *PHOTO: TOM BORDEAUX*

How Much Solar Hot Water Do I Need?

The typical U.S. home uses about 20 gallons (76 liters) of hot water per person, per day for the first two people in the home. After the first two, figure about 15 gallons (57 liters) for each additional person per day. Assuming your nice hot water will need to be heated from 45° to 125°F (7° to 52°C)—a typical rise—you'll invest about 835 Btus into each gallon, or roughly 15,000 Btus per person per day. If there are four people in your home, your daily hot water energy consumption will be around 60,000 Btu.

Solar collectors are rated by how many Btu per day they'll deliver. What could be simpler? But not so fast. Collector ratings are developed under laboratory conditions, and as we've already (hopefully) learned, it's not a laboratory out there in the real world. In most cases your solar collector is not going to deliver as much energy as the FSEC or SRCC ratings would have us believe. If you're in the Northeast, performance may be as much as 25% less. The collector ratings are useful to compare one collector against another in an apples to apples comparison. Your actual "mileage" may vary.

Solar energy, or irradiation, varies seasonally and regionally. In most of North America a solar hot water system capable of delivering 75%–80% of your hot water needs is ideal. Squeezing out that last 20% requires increasing amounts of hardware with diminishing returns. It's not worth chasing. As a rule of thumb, a good-quality glazed flat plate collector should be SRCC rated to deliver about 1,000 Btu per

Please Note...

Recycled old flat plate panels without an SRCC rating do not qualify for the federal tax credit.

Rebates or Tax Credits for Solar Thermal

A 30% federal residential tax credit is available for any solar hot water system producing domestic hot water. Heating your pools or hot tub does not qualify. (Wow, common sense from the Feds!) Since these systems average between $3,000 and $10,000, this tax credit is significant. In addition, many individual states or cities offer rebates or credits. Please check your local incentives at the Database of State Incentives for Renewables and Efficiency at *www.dsireusa.org*. You're allowed to double-dip these federal credits, so if you install a PV system and a solar hot water system in the same year, there's no loss of available credit.

Comparing Apples to Apples

Every manufacturer is eager to tell you how their product is better than the other guy's product. Fortunately, there are rigorous standardized performance tests designed to reveal just how good a product is. Specifically, solar hot water systems are certified by either the SRCC or FSEC *(see page 155 for web sites).* If you are considering a product that doesn't yet have a certification, don't despair. It takes up to two years to get an SRCC rating, so ask the manufacturer what their plans are for certification.

square foot of surface area on a good day. A common 4-foot by 8-foot collector has 32 square feet. You should expect an SRCC rating at class C (36°F / 2°C rise) of 30,000 to 35,000 Btu per day. So a small household of two people should usually do well with a single 4 x 8 or 4 x 10 collector. If you've got more bodies or not a lot of sunshine, you'll need more square feet of collector. System sizing is more art than science, as output depends on regional climate, orientation of the collector(s), and how aggressively you want to go after those "free" Btus. If you're willing to settle for a smaller solar percentage, your Btu return per dollar invested will be higher, but you'll spend more on backup water heating energy. This is a good subject to discuss with your local installer who has a better working knowledge of local solar conditions and knows from experience what can be expected from a particular system.

Choosing the Best Type of System

What do you need to know to pick a good system? The first and most important deciding factor is your climate. As we all know, water freezes regularly in most of North America. And when it does, it expands about 10%. This is great for ice skating and the automotive body shop business, but it's rough on plumbing. We're going to do our best to make sure your plumbing avoids this fate. Solar hot water systems come in two primary flavors, those for freezing climates and those for non-freezing climates. For those of you who aren't sure if you live in a freezing climate, here's the test: if it gets into the 20s(°F), and stays there for longer than 24 hours, and it does this more than once every 10 – 15 years, you're in a freezing climate. Solar hot water systems for non-freezing climates can (usually) tolerate an overnight freeze, but not one that lasts more than a day.

Solar Hot Water Systems for Non-Freezing Climates

If you're blessed with a non-freezing climate, then a simple batch-type solar hot water collector is your ticket to hot water bliss. But because batch heaters typically involve 40 or 50 gallons of water (at 8.3 pounds per gallon), weight is a consideration. Situating it on the ground propped up against a south-facing wall might be smarter than a roof mount. If roof-mounting is your best choice to avoid shading, then you may need to double up a couple of your rafters to help support the weight, which can be around 500–600 pounds (227–273 kg) for a batch heater.

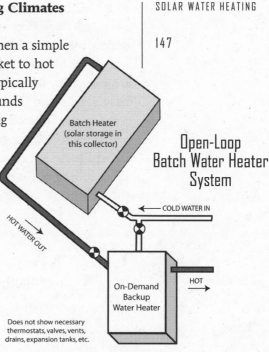

Open-Loop Batch Water Heater System

Batch Heater (solar storage in this collector)

HOT WATER OUT

COLD WATER IN

On-Demand Backup Water Heater

HOT

Does not show necessary thermostats, valves, vents, drains, expansion tanks, etc.

There are a couple things to look for in a batch-type heater. The glazing should be double-layered like a thermo-pane window for better insulation. Some manufacturers use a layer of plastic film for the second layer to save weight. The result is the same: an air space that helps hold in heat. The cold inlet should be at the bottom of the tank and the hot outlet should be at the top. The water will stratify in the tank with the warmest layers at the top, and that's the water we want to deliver to the house first.

Solar Hot Water Systems for Freezing Climates

There are two ways to deal with freezing. You can buy one of the less expensive solar hot water systems that are designed for a non-freezing climate and make a solemn promise to yourself that you will absolutely, positively shut down and drain your system before the

If you're blessed with a non-freezing climate, then a simple batch-type solar hot water collector is your ticket to hot water bliss.

first hard freeze every year. Then you reactivate it in the spring. This isn't really recommended. Besides not reaping solar-heated water during that time, guessing (aka forecasting) the weather has never been one of humanity's strong points. But if you're in a marginal climate that threatens freezes for only a month or two every year it could be workable, and the difference in system cost makes it attractive.

A better solution might be to invest in a system that uses some sort of non-freezing fluid in all the exposed parts. These are called "closed loop" systems. There's a plumbing loop that goes out to the solar collector, picks up heat, then comes back in and is routed through a heat exchanger where it gives up that heat to your domestic hot water. It's a continuous loop that is completely separate from your household water system, so we can fill this loop with some kind of anti-freeze solution and be absolutely assured that it won't freeze and cause any plumbing damage, no matter how cold it gets.

There are a couple of other plumbing schemes for freezing climates that bear mention. Drain-down and drain-back systems were widely used in America's big wild fling with all things solar following the Arab oil embargo in the 1970s and '80s. These systems depend on the collection fluid—plain water—draining out of the collector at night. Reliability and

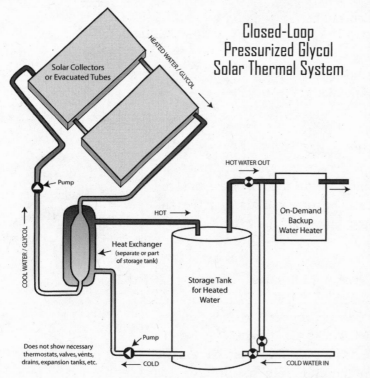

Closed-Loop Pressurized Glycol Solar Thermal System

Solar Collectors or Evacuated Tubes

HEATED WATER / GLYCOL

Pump

COOL WATER / GLYCOL

Heat Exchanger
(separate or part
of storage tank)

HOT

Storage Tank
for Heated
Water

Does not show necessary
thermostats, valves, vents,
drains, expansion tanks, etc.

Pump

COLD

HOT WATER OUT

On-Demand
Backup
Water Heater

COLD WATER IN

long-life expectancy weren't their strong points. I personally converted many of these failed systems to closed-loop glycol systems in the late '80s and early '90s. And I work in the (almost) non-freezing climate of coastal northern California. For this reason we suggest you look long and hard at all your options before deciding on a drain-down or drain-back system.

Circulating Heated Water

Closed-loop systems need a pump to circulate the heat-collecting fluid out to the solar collector and then back through the heat exchanger. It's obviously a good idea if that pump only runs when there's heat to collect. So your closed-loop system will have one of two control types. Standard controllers use a pair of temperature sensors, one at the storage tank, and one at the collector. If the collector gets around 10° to 14°F (-12° to -10°C) hotter than the storage tank, the controller turns on the pump and will run until there's about a two degree difference, then shut off. It's not worth continuing to run the pump with only a tiny difference in temperature. Depending on the heat exchanger design and placement (more on that later), it may be necessary to run a second pump to circulate the domestic water through the heat exchanger whenever the closed-loop pump is running.

This kind of temperature-sensing electronic control system has the advantage of perfect control. It can be programmed to turn on and off at just the right temperature points for optimum performance.

Anti-Freeze in My Hot Water System?

Isn't anti-freeze the stuff in my car radiator that kills wildlife and the family dog if they drink it? The green stuff in your car's radiator is ethylene glycol, and yes, it's a deadly poison. Worse, it tastes good (take my word for that please). The anti-freeze used by the solar hot water industry is propylene glycol, a close relative of ethylene glycol, but not poisonous. In fact, propylene glycol is FDA-approved as a food additive. Tastes like cream soda. Also, inside the heat exchanger there are two layers of metal between your domestic water and the closed loop. Manufacturer's use "double-wall" heat exchangers just to be really safe, in case some gung-ho but misinformed solar enthusiast really does put automotive anti-freeze in his solar hot water system.

It has the disadvantage of requiring remote sensors, one of them outdoors. Sensor, or wiring and connector failures, are the most common service call on these systems. Mice chew through wires, weather corrodes connectors, sensors die. In theory, this control system is great. But in the real world, it's proven to have some trouble points that should be adequately addressed by your installer.

Personally, I'm a real fan of the other control system…PV control. With this type of system, a small 10- to 20-watt PV module is directly wired to a DC circulation pump in the closed loop. The PV module and the pump are closely matched, so that if there's enough sunlight to get the pump going, there's enough sunlight to produce harvestable heat. The brighter the sun shines, the more heat is collected, and the faster the pump runs. No sensors are required, and the wire runs are very short. Simple!

PV controls have the advantage of simplicity and reliability. But there are disadvantages. They may, for instance, start pumping when the collector is colder than the storage tank, especially on clear cold mornings. And they may continue pumping later in the day when the delivered fluid is cooler than the storage tank.

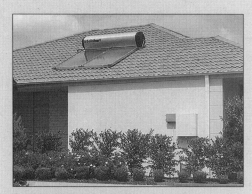

SolaHart's thermosiphon roof-mounted system.
PHOTO: SOLAHART

Look Ma! No Pumps!

Solahart, Sun Earth, and other manufacturers make a series of innovative flat-plate solar hot water systems that depend entirely on thermosiphon action. No controllers, no sensors, no pumps, no moving parts to wear out. The hot water in the collector rises into the storage tank mounted above. These collector-with-tank systems are available with open- or closed-loop collector plumbing for temperate or freezing climates. On-demand, tankless heaters are a perfect backup to ensure you have hot water even on cloudy days.

Both these problems can be solved if the domestic hot water loop and separate heat exhanger are set up as a passive thermosiphon, instead of being pumped. "As a what?" I can hear you saying. A passive thermosiphon works by installing the separate heat exchanger 3 or 4 feet (.9 – 1.2 m) lower than the top of your storage tank. As the domestic hot water is heated by the heat exchanger, it wants to rise. As it rises into the storage tank, cooler water is pulled off the bottom of the tank into the heat exchanger, where it in turn warms and rises. This heat-activated loop keeps circulating as long as the fluid from your solar collectors is hotter than the domestic water. A special, lightweight, one-way valve in this loop insures it can only circulate in the upward heat-collection direction. If the solar-collection fluid is cooler than the domestic water, which can happen early or late in the day, the one-way valve prevents the thermosiphon from running backwards. It's a simple and foolproof system if you choose a separate heat exchanger and a stand-alone solar storage tank.

Where to Store that Hot Water

Your solar hot water system needs someplace to store the heat it collects during the day. For batch systems this is easy; the collector is also the storage tank. Hot water is held in the collector/tank until needed. For flat-plate or evacuated-tube systems, however, we'll pump the collected heat to a hot-water storage tank. For a small one- or two-person system, this could be your previously existing 40- or 50-gallon hot water tank, but more commonly it's going to be a new 60- to 120-gallon (227- to 455-liter) solar tank. This new storage tank will be plumbed inline ahead of your existing hot water heater. In this way your existing heater gets pre-heated water delivered to it and becomes the backup heater, adding a few degrees to the pre-warmed solar water if needed.

Water Storage Tanks for Solar Water
A typical 50- to 60-gallon storage tank is about 24 inches in diameter by 60 inches high (61 cm x 152 cm). Other sizes are available, from 40 to 120 gallons (150 to 455 liters).

Chances are your storage tank will be a specially designed solar tank with a built-in heat exchanger. This has pretty much become standard practice in the industry. It is the least expensive way to go, and it will certainly save your installer some plumbing work. But with this setup you run the risk that someday the storage tank will begin to rust out and start leaking—with a perfectly good (and very pricey) heat exchanger locked up inside of it. For this reason some home-owners insist on an external heat exchanger plumbed inline, en route

Use Tankless Water Heaters for Efficient Backup

Using your existing tank-type heater as the backup to solar-heated water works fine, but for the most efficient backup water heating system, a tankless (on-demand) water heater is ideal. Tankless heaters only run when someone turns on the faucet, and they'll only run as much as needed to bring the already pre-warmed water from the solar system up to temperature.

Solar pre-heating and tankless water heaters are a match made in heaven. Each technology complements and completes the other. This is particularly true if you're using a gas-fired tank-type water heater (either propane or natural gas). Because of the flue running up through the middle of the tank, these heaters are always losing heat. A tankless heater is the only way to beat this 24/7 constant heat loss. So go ahead and use your tank-type heater for backup now, and promise yourself that you'll upgrade to tankless when the old heater starts to leak in a few years.

Just make sure to buy a model designed to receive preheated water from your solar storage tank, such as the Aquastar "S" model, Takagi and Paloma models (all gas-fired), as well as Stiebel-Eltron (electric).

The Takagi T-K3 accepts solar-heated water up to 160°F (71°C) and the burner modulates from 11,000 to 199,000 Btu. It doesn't take up much space at 21x14x9 inches (53x36x23 cm). The minimum flow rate of water to activate the burner is 0.5 gallons/minute. *PHOTO: TAKAGI*

Adventures in Plumbing: Rex's Solar Hot Water System

Plumbing has never been my cup of tea. All my memories of plumbing—most of them dealing with frozen livestock waterers in the dead of winter—are bad. So when LaVonne and I hand-hewed our off-grid log house a few years ago, the plumbing was one of the very few things for which we were willing to shell out money to an outside contractor. But in the fall of 2007, when the price of propane was bouncing around the $2.50-per-gallon mark, I decided—with confidence-boosting assurances from a local installer who consulted on the job—to temporarily don my plumber's hat and install our own solar domestic hot-water system.

Taken as a whole, it wasn't all that bad. No lingering injuries or major screw-ups at any rate. Still, I can't say I was having much fun the day I found myself clinging by my toes to our 9/12-pitch roof, sweat-soldering copper pipes with a propane torch as the wind gusted from the south at 30 mph.

We chose to go with a system comprising 30 Thermomax evacuated tubes, even though we had a few of the bulky Carter-era flat-plate collectors. Our decision added measurably to the cost of the system (which is bad), but the tubes—each 2½ inches by 72 inches—and the manifold were light enough that we could install them on our high, steep roof without

having to hire an expensive boom truck or a football team (which is good). And we could also expect much better performance from the compact array of tubes on overcast days and in winter.

Technically, the system we installed is a closed-loop pressurized-glycol system. An adjustable-speed AC pump wired into our home's solar-electric system ferries freeze-proof glycol in a continuous loop between the tube manifold and an 80-gallon Rheem Solaraide solar tank with a built-in heat exchanger, neatly stuffed into a crowded mechanical room 25 feet below the tube array. The pump marches to the tune of a differential thermostat that measures the difference in temperature between the water in the tank and the heating-fluid in the manifold. Whenever the latter is 14 degrees warmer than the former, the pump runs. A series of three isolation valves allows us to bypass the backup heater during the warm months, and during sunny periods in winter.

Our total material cost for the system was around $6,500. How well does it work? As I sit writing this, the wind chill outside is -6°F while the heating fluid leaving the manifold is a comfy 112 degrees.

I'd really hate to ask for more than that.

— *Rex Ewing*

to the storage tank. Then, if the storage tank does eventually begin to leak, it can be replaced with a top-quality, off-the-shelf, electric water heater. Minus, of course, the expensive heat exchanger. Just don't hook up the electric.

Other Installation Considerations

Which Way to Aim the Collectors   Just like with solar electric systems, in North America you ideally want to orient your solar thermal collector due south, and at an angle about equal to your latitude, which is 30° to 45° for most of us. But if south just isn't a doable direction on your house due to roof angles, or shading, or a backward-thinking homeowner association, any orientation from due east to due west is workable. You'll only lose 12% of your potential performance facing east or west, and that quickly improves as you edge around toward southeast or southwest.

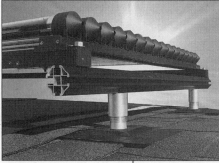

The Zilla mounting system shown with Next Gen Energy's thermal tube collectors.
PHOTO: NEXT GEN ENERGY

It's the same story with the tilt angle of your collectors. Tilting at the same angle as your latitude will deliver the best performance year round, but most of us have roofs with 18° to 25° angles (4/12 to 5/12 pitch), rather than the 30° to 45° angle (that's a 7/12 to 12/12 pitch) you would need to match your latitude. It's a clear case of functionality versus appearance. Flush with the roof or raised to latitude? While I could certainly voice an opinion in the name of eye appeal, it's really your call.

Distance and Shading   We need to give some consideration to several other factors before installation can begin. Distance is more important with solar thermal collectors than with solar-electric modules. They don't transfer those hard-won Btus as easily as solar-electric modules move electrons, and plumbing is far more expensive than wiring. So your thermal collectors should be as close as practical to the storage tank or backup heater. If you can keep it within 50 feet (15 m) that's great. If you have to run 100 feet (30 m) or further, system performance is going to suffer.

A little shading is no big deal with thermal collectors. You'll lose solar gain from the shaded portion, but unlike solar electric modules, all the Btus from the rest of the collector won't rush over and simply disappear into the shade. (Really, this is what happens with shade on solar-electric modules!) So if you've got limited roof space and are trying to fit both solar electric and solar thermal onto it, it's good planning to put the thermal collectors where you know there's going to be some limited shading and give the PV modules full sun.

Freeze-Proofing   If you're in a "non-freezing" climate and using collection fluid that's capable of freezing (it's called water), then freeze protection needs to be high on your priority list. Unless you live in Hawaii, all climates freeze occasionally. Finding your freeze protection was insufficient after the pipe in the attic has burst, dumping 2,000 gallons of water—which soaked through the insulation and caused the drywall ceiling to collapse—is not anybody's idea of a good time...been there, done that...yuck!

Batch collectors are fairly immune to overnight cold snaps because they're filled with around 500 pounds (1,100 kg) of water. But the supply and return plumbing is still vulnerable. Keep the plumbing runs in heated space as much as possible, and don't skimp on the pipe insulation—buy the highest-quality stuff you

For More Information on Solar Water Heating

FSEC (Florida Solar Energy Center)
www.fsec.ucf.edu/en/consumer/solar_hot_water/

SEIA (Solar Energy Industries Association)
www.seia.org

SRCC (Solar Rating and Certification Corporation)
www.solar-rating.org

can get. Besides possibly preventing a frozen pipe, it will greatly improve performance on cool days.

Flat plate collectors freeze with incredible ease and need to either be completely drained or (better yet) hooked up to an anti-freeze, closed-loop system. There are some specialized poly-pipe systems specifically designed for solar hot water systems that will tolerate freezing without breaking. If you're in one of those marginal almost-non-freezing zones your local solar thermal folks will know about them, and hopefully recommend them.

Doug's Cheap Solar Hot Water

It's not that I'm a cheapskate, but why buy a Lexus when a Camry gets you there just as well (or in my case, a Prius)? My solar hot water system compared to Rex's is a good study in contrasts. His is a more-expensive, high-performance system that shrugs off sub-zero

temperatures. It grabs any Btus to be had and just never lets them get away. In my milder (mostly-non-freezing) northern California coastal climate, I use an inexpensive batch-type water heater that delivers about 80% of my hot water needs per year. It's not as aggressive about grabbing and hanging onto those Btus, so I sacrifice winter performance from mid-November through January. But from March to October we can take back-to-back morning showers without kicking on the backup heater.

My ProgressivTube solar collector is 4-foot by 8-foot and consists of a series of large diameter copper tubes connected in a zigzag pattern. The tubes have a selective-coating so they're about four times better at absorbing heat than radiating it. They're protected

Listen to the Locals

If you can find a local solar thermal installer who's been in business for several years, take seriously his or her advice on what to buy and how it install it. Solar hot water systems are climate-sensitive. If your local pro has a brand and type of system he or she prefers to install, it's probably for good reasons other than profit. Quality lasts. Good installers should have few service calls, and a long list of satisfied customers you're welcome to call. In this business the best referrals, for both the dealer and the homeowner, come from neighbors.

www.FindSolar.com is a good source for locating local installers.

---

by 4-inches of rigid foam on the back and sides, and a thermo-pane air gap on top. It holds 40 gallons of water, with the cold entering the bottom and the warmest water in each tube stratifying up into the next tube, with the hot outlet at the top. The collector is plumbed inline so its hottest output is the backup water heater's cold inlet. There are bypass valves at the backup tank-type electric water heater, so we can choose solar pre-heating or just run off the backup heater. There's also a pair of drain valves just in case we get a real serious cold snap where it stays below freezing for more than a day. (It happens once every 10 to 20 years around here, and boy does it keep the plumbers busy!)

The collector, with all its mass of water and heavy insulation, isn't threatened by overnight cold snaps, but the supply and return plumbing is. So I was careful to keep the plumbing runs between the warm insulated house and the collector as short—and well-insulated—as possible. Both pipes are under two-feet in length.

Since weight is a concern with batch heaters (when filled, mine tips the scale at over 600 pounds), our system is sitting on two doubled-up pairs of 3 x 8's. Probably overkill, but better safe than saggy.

What did it cost me? The collector, mounting, and other plumbing parts came to under $3,000. Then I invested about 12 hours of sweat-equity in plumbing and insulating. In return I get a lifetime of free hot water. Sweet!
— *Doug Pratt*

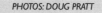

PHOTOS: DOUG PRATT

Too Hot by Half?

In the summertime it is not uncommon for the collection fluid in the solar collectors—and therefore the solar storage tank—to reach temperatures beyond the boiling point, creating a potentially dangerous situation. While a mixing valve (aka tempering valve) will automatically mix enough cold water with the hot to keep the domestic water en route to your taps a constant temperature, it can be a bit more complicated for the rest of system. Be sure to ask your installer how he/she plans to address this concern.

Financial Help and Tax Credits

There's definitely money in the form of tax credits and grants available for homeowners using solar energy and conservation. Details are still being worked out, but if this works anything like previous solar tax credits and grants, you'll need to show a receipt for purchase, an invoice for professional installation, and a building permit. This is another good reason to get involved with your local solar company.

If you want to keep costs down and just can't keep your hands off the hardware, make it clear that you want to be their helper on the job and be as hands-on as they'll allow. Most contractors are happy to work with homeowners. Having a licensed contractor pulling the permit and having the job inspected insures a quality installation and no trouble with those pesky rebate/tax credit bureaucrats.

And as we mentioned on page 145, a 30% federal residential tax credit is available for any solar hot water system producing domestic hot water. ❖

Installing a flat plate collector on a home's roof for a solar hot water system. *PHOTO: STEVEN C. SPENCER, FLORIDA SOLAR ENERGY CENTER*

CHAPTER FIFTEEN

Geothermal: Capturing the Heat Beneath Your Feet

Imagine for a moment that you hear someone talking about a magic battery capable of collecting the oppressive warmth of the summer sun and storing it until winter, when you really need it to heat your house. By using the magic battery, you are told, you will never have to burn another gallon of propane, natural gas or fuel oil to keep your house warm. Too good to be true? Not at all. In fact, the magic battery is nothing more otherworldly than the ground beneath your feet, and the means used to move the heat back and forth is the same clever science that makes refrigerators and air conditioners possible: the common heat exchanger.

> You don't need a geyser in your yard to heat your house with Earth's warmth. In most cases, nature's balance keeps the ground a constant temperature perfect for heating and cooling.

The Principle of the Heat Exchanger

Most of us know the earth stays a fairly constant temperature several feet below the surface. That's because it takes so long for the ground to gain or lose heat, by the time it begins to react to one season, the next season is already underway. So if you dig down 5 to 8 feet (1.5 – 2.4 m), the soil will be a fairly constant 45°– 50°F (7°– 10°C) in the northern latitudes, and 50°– 70°F (10°– 21°C) in the south.

This is all easy enough to understand. But now it's about to

Minnesota Comfort

The Willenbring's 4,300 square-foot home/garage in Minnesota is heated and cooled with an open-loop geothermal system that uses a series of twelve water wells. In addition, a desuper water heater provides essentially free hot water in summer and very cheap hot water the rest of the year. By taking advantage of off-peak electric rates, their average monthly heating/cooling bill is under $100. At prevailing propane prices, they calculated the payback of this system was about 3 years. Not bad, for a family of six in the far north.

get counterintuitive, because I'm going to tell you that you can heat your home cheaply and conveniently by extracting heat from soil that is only 15 or 20 degrees (F) above freezing. How? By circulating, through buried tubing, water that is even colder than the earth itself. Since heat always flows toward where it's colder, the water—mixed with an environmentally-safe anti-freeze solution—picks up heat from the ground as it circulates. Thus it will be warmer when it returns to your house than when it left it.

The water is still cold, of course; no warmer than the earth it came from. So it has to give up its heat to something even colder—a refrigerant, such as liquid freon. Now we're getting somewhere, though it may not seem like it, yet. Inside a heat pump unit, the very cold freon circulates in a double coil with the earth-warmed water, absorbing its heat and making the water cold again, relatively speaking. But even though the freon has absorbed most of the heat the water gained from the soil (becoming a low-pressure gas in the process), it's still no warmer than the ground outside. And we need to make it hot; hotter even than the air inside our house. How? By compressing it to a very high pressure. This concentrates the gas and raises its temperature to approximately 165°F (74°C). Then, by running the hot high-pressure gas through a second heat exchanger (either an air duct coil or a hot water tank), the heat is given up into your house. This causes the hot gas to cool, become liquid again, run through an expansion valve, and return to a cold, low-pressure state, ready for the next go-round. In summer the process is simply reversed: the excess heat inside your home is returned to the soil.

With three separate phases of heat exchange (ground to water; water to freon; freon to air or water), the use of a ground-source (geothermal) heat pump to heat and cool your home may not, at first blush, appear to be a very efficient process. But in fact it is extremely efficient—right in the neighborhood of 400%. This means that the ground surrenders four units of heat energy for every one unit of electrical energy you use to extract it. It's almost like stealing.

Types of Geothermal Systems

There are two basic types of underground pipe systems—open loop and closed loop—with several variations. The one you use will depend on where you live, how much land you have, and the characteristics of the soil and ground water.

Closed-Loop Systems   In a horizontal closed-loop system, loops of special heat-conducting polyethylene pipe are typically buried in a series of 3-foot-wide (1 m) trenches 3 to 8 feet (1–2.4 m) below the surface. The length of each trench depends on the amount of moisture in the

Federal Tax Credit
Geothermal systems are eligible for a 30% federal tax credit, with no cap on the credit.

Residential Geo-Exchange System

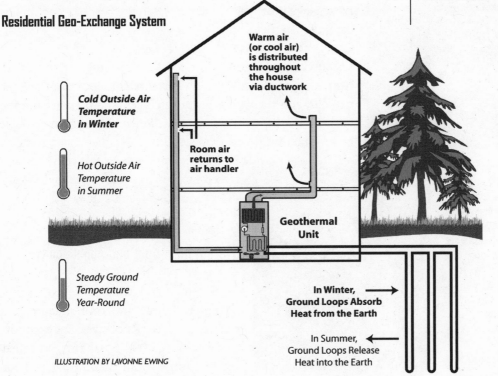

Warm air (or cool air) is distributed throughout the house via ductwork

Cold Outside Air Temperature in Winter

Hot Outside Air Temperature in Summer

Steady Ground Temperature Year-Round

Room air returns to air handler

Geothermal Unit

In Winter, Ground Loops Absorb Heat from the Earth

In Summer, Ground Loops Release Heat into the Earth

ILLUSTRATION BY LAVONNE EWING

soil, ranging from around 100 feet (30 m) for saturated soil to 200 feet (61 m) per trench for drier soil.

Since many of us simply don't have enough land for such a sprawling system, another option is a vertical closed-loop system, in which loops of ¾-inch high-density polyethylene pipe are set in concrete in a series of 4½-inch (11.4 cm) holes. The holes are bored to a depth of 150 – 450 feet (46 – 137 m), and placed 10 – 15 feet (3 – 4.6 m) apart. All the separate pipes converge at a manifold, where they are joined into two pipes—one in, one out.

A third incarnation of the closed-loop system is the pond loop. As the name implies, the tubing is floated over a body of water, then sunk to the bottom. If you're lucky enough to have a lake nearby, it could spare you the cost of trenching or drilling.

Open-Loop Systems   In open-loop systems, ground water from a series of wells is used. Water is pumped out of one set of wells, run through a heat exchanger, and is then pumped back into a different set of wells. Since water conducts heat better than dirt, open-loop systems are very efficient and can be less expensive to install than vertical closed loops if you don't have to drill very deep for a good water supply.

System Size, Cost and Applicability

In a temperate climate, a geothermal system will heat and cool roughly 750 square feet (70 m²) of space per each ton of capacity (12,000 Btu/hour). Installed systems range from $3,000 per ton for

Specialized high-speed, lightweight drill rigs are often used to punch the vertical bore holes. Here you see a heat exchanger loop being inserted into a finished bore hole. (see page 164 for more details).
PHOTOS: DOUG PRATT

horizontal closed-loop systems in ideal soil, to $5,000 per ton for vertical closed-loop systems. If, on the other hand, you have adequate groundwater flow, vertical open-loop systems can save you money.

Are you building a new house? If so, the increase in your mortgage payment from choosing a ground-source heat-pump system over a conventional system will be more than offset by the savings on your utility bill, since a geothermal system can be up to three times more efficient. If you instead plan to retrofit an existing house, the payback will take 10 – 15 years, depending on numerous factors. The good news is, the heat pump equipment should last 20 – 30 years with little maintenance, while the underground tubing will perform trouble-free for at least 100 years.

Practically speaking, ground-source heat pumps work best with forced-air heating systems, but can also work well with radiant-floor heating systems. When adapted to an existing forced-air setup, the system efficiency can be greatly augmented by sealing gaps and holes in the ducts, adding return-air ducts, and setting up multiple zones. With the addition of a desuper water heater, you can enjoy virtually free hot water in summer and more efficient water heating in winter.

When installed as a hot-water heating system for radiant in-floor heating, a geothermal system can only heat water to 110° – 120°F (43° – 49°C), so for most homes it will require some amount of boiler-heated water on really cold days, though overall it should handle 80% – 90% of the heating chores over the course of a winter. Summer cooling with ground-source hot-water systems is accomplished by reversing the process. The heat pump unit produces cold water which is circulated to fan coil units which blow air past the cool coils. The cool air is then distributed through a separate duct system.

Want to do something good for both the planet and your pocketbook? According to the EPA, geothermal systems are the most energy-efficient, environmentally clean, and cost-effective space conditioning systems available. ❖

For More Information about Geothermal

To learn more about geothermal heating and cooling, or to find a certified installer near you, visit the GeoExchange website at: **www.geoexchange.org**

For a comparison of **heating fuel costs**, see page 51.

Living with a Geothermal Heating / Cooling System

As we were planning our new passive-solar home in 2006, we were already looking seriously at geothermal-based heating and cooling. Although these systems tend to be expensive to install, they cost a third less to operate than any other heating or cooling system. When we found that our local small-town heating/cooling contractor was now a fully-trained dealer/installer for these systems, that sealed the deal. (We like to spend money locally whenever possible.)

The idea of a geothermal system was attractive from the get-go for its efficiency and low cost of operation, but the fact that I can actually make the fuel—electricity—for a geothermal heat pump was especially compelling, since I obviously can't make the fuel for a propane-fired heater. So heating and cooling costs will be one less thing to worry about if my retirement ever comes.

Summer cooling, which our inland California 100°(F) summers demand, requires a forced air system, and much as my tootsies would have enjoyed a radiant floor, we just couldn't justify the quite significant extra cost. Besides, radiant floors are best suited for climates that have constant heating loads. Tons of concrete floor mass don't turn on and off at the flip of a switch. Our passive-solar home would need some heating help on clear cold nights, but very little on the sunny day that would follow. How do you tell all those Btus in the floor to just stay put for 12 hours?

Our system and installation cost about $20,000; $12,000 of that for the pair of 250-foot holes in the backyard. Due to terrain and property lines, a trench-laid plumbing loop would have required some zigzagging, and would have cost as much or more. Because our home is small (1,450 sq. ft.), highly insulated, and passive solar, we were able to get away with only two boreholes. The boring, which was done with a specialized high-speed rig, and installation of the closed-loop plumbing only took a couple days. (*See photos on page 162.*)

How do we like living with geothermal? As I write this we've been in the house for about 16 months, including two winters. The most noticeable trait is how unnoticeable it is. The system runs almost silently. Unless you're in the garage, it's practically impossible to tell if the heater or AC is running. Set the thermostat higher, the house gets warmer; set it lower, the house gets cooler. The only time I'm aware of the heater is if I'm seated under a vent when the heater starts up.

Power use is about 2,000 watts, with a rated delivery of 24,000 Btu. To save you the math, that's a little over 7,000 watts output. Heat pumps don't make the energy, they just move it from one place to another. Hence, you seem to get more energy out than you put in. Pretty slick. And I won't complain about my $100 electric bill since that's for the entire past year and I live in an all-electric house.

— *Doug Pratt*

Postscript

In 1966, in a little-known book titled *Building Blocks of the Universe,* Isaac Asimov wrote with customary prescience, "Recently, solar batteries have been designed which can produce an electric current when exposed to sunlight. So far, such batteries are only laboratory curiosities, but the day may come when they will be an important source of power for mankind."

An insightful call from a great science writer. But not even a visionary like Asimov could have guessed back then the extent to which the science and applicability of solar "batteries"—as solar cells were originally called—could have advanced in the course of a few decades. From laboratory curiosity, to the space program and a few limited terrestrial applications, and finally to the ubiquitous solar arrays powering millions of homes and villages worldwide, solar electricity has at last come of age.

Solar- and wind-electric applications, glitzy as they may be, are hardly alone in this exciting planet-wide endeavor. The sun—as we've always known but have only recently methodically acknowledged—has far more to offer. Solar water heating technologies, with new high-tech upgrades, are back in force and here to stay. And geothermal energy has waited patiently underground for millennia to be mated with today's state-of-the-art, high-efficiency ground-source heat pumps. There is simply no longer any reason for homeowners to be captive promulgators of the ill-conceived energy practices of the twentieth century.

But, as widespread and accessible as solar technology may be today, what we see around us is merely the beginning. If we're as insightful as we like to think we are, twenty years down the road solar technologies—and their sidekick, wind—will have made great strides toward replacing coal-fired and nuclear power plants. Cars that once ran on gasoline will run on non-polluting sources of electricity, or on hydrogen cleaved from water or biomass by solar and wind applications. Every new home will be designed to make the most of its natural setting for lighting, heating and cooling, and all the extra energy needed to power it and keep its occupants in the comfort zone will be derived from environmentally responsible sources.

In time, the absurdity of the self-destructive conventions we cling to in the name of progress will become universally evident. Future generations will look back on our current energy production practices with a mixture of bemusement and revulsion, much as we now look back at the barbaric and superstitious medical procedures of the Middle Ages.

Until then, we can each help to ensure a long, bright future for a really spectacular planet by spurning the unchecked burning of fossil fuels, one renewable energy application at a time. So, what are you waiting for?

PHOTO: LAVONNE EWING

Got Sun? Go Solar!

Acknowledgments

Any book, even one as seemingly simple as this one, is the work of far more people than the authors. Sure, we write all the words and get our names on the cover in big letters, but without the assistance of numerous others who toiled to turn a raw manuscript into a bona fide book, the result wouldn't be anything you'd want to read. So it's only fair we take the time to thank everyone who went out of their way to make us look good.

First and foremost, we'd like to thank LaVonne Ewing, who humbled us both by working harder and complaining less than either of us glory hogs. Without her organizational skills and design talents *Got Sun? Go Solar* would never have sprouted wings. As an editor, she can nurse a crippled phrase to a perfect state of health, or axe a misbegotten one so painlessly it seems an act of mercy. Thanks, LaVonne—you're the best.

We'd also like to thank all the homeowners who provided photographs of their grid-tied homes, and happily agreed to being a part of this book. You are an inspiration to us all.

Home-based solar and wind energy systems would still occupy the realm of science fiction without the hundreds of companies and corporations who believed so strongly in the concept of renewable energy that they were willing to risk it all by engineering and manufacturing the state-of-the-art components needed to harness the power of the sun and wind. Many provided photographs or hard-to-find information that helped to make this book what it is. We thank you.

A special thanks to Christopher Freitas from Outback Power Systems for sharing his insights on the possible future of residential renewable energy.

And finally, we'd like to thank the diligent women and men at the National Renewable Energy Laboratory. You saw the grail when it was invisible to the rest of us. Without you, this would be a really short and boring book.

Tips on Appliances and Energy Conservation

Use compact fluorescent light bulbs or new LED light bulbs...they add up to big energy savings. For example, if 6 bulbs are on for 5 hours a day: 60-watt incandescent bulbs will use 1,800 watt hours per day; 13-watt compact fluorescent bulbs will use only 390 watt hours. LED lights will barely consume any wattage.

Use the on/off switch. Like Mom always told you, turn off the lights when you aren't using them. Put infrequently used lights on motion sensors or wind-up timers. Outdoor floodlights are also perfect for motion sensors.

Shut down your computer when you're not using it. Besides, it's healthy to reboot regularly. Put all computer peripherals on a power strip and turn it off when done.

Low-usage, high energy appliances (hair dryers, microwaves, coffee makers, etc.) are not much of a problem since they draw very little power when averaged out over time. You can also choose not to use them.

Invest in a new **refrigerator and/or freezer**. You'll be amazed at how much more energy-efficient they are. A new Energy Star model uses half the energy of a 1993 fridge. Do your research on *www.energystar.gov* before buying and always read those yellow tags.

If you want to cook when the grid is down, and you don't have battery backup, buy a gas **oven** with spark ignition instead of the typical glow bar. Glow bars use 300–400 watts ALL the time your oven is on. Peerless-Premier is one brand that has done away with the glow bar.

To conserve energy and water when washing clothes, a **front-loading clothes washer** is a must, as is a gas-fired clothes dryer. Better yet, use a clothesline or indoor rack for drying.

On-demand (instant) water heaters, either gas or electric models, use 20% – 40% less energy because they only work when someone turns on the hot water faucet. They also last 30–40 years, reducing landfill and resource waste.

One watt delivered for one hour = **one watt-hour** | 1,000 watt-hours = one **kWh**
amps x volts = watts (*2 amps x 120 volts = 240 watts*)

Energy Consumption of Appliances

Using a WATTS-UP? Meter, we measured the following appliances.

Appliance	Continuous Draw (Watts)
Computer, desktop (CPU only)	90
Computer, laptop	24
17" Computer Monitor	100
22" LCD (flat screen) Monitor	30
HP LaserJet Printer (in use)	600
HP Inkjet Printer (in use)	15
Microwave (full power)	1,400
Coffee Maker	900
Toaster, 2-slice	835
Amana Range (propane): Burners	0
Oven (with glow bar; when heating)	380
Electric Range (small/large burner)	1,250/2,100
Blender	350
Mixer	120
Slow Cooker (high/low)	240/180
19" Magnavox LCD Television/DVD player	40
26" JVC Television (LCD)	104
27" Television	120
50" LCD Television	175
Stereo System	25
Stereo, small portable	10
Bose iPod Docking Station / Stereo	less than 1
Vacuum, Oreck	410
Vacuum, Dirt Devil Upright (10 amp)	1100
Table-top Fountain	5
Sewing Machine (Bernina)	70
Christmas Tree Light String	
- 200 mini lights	68
- 200 new LED lights	6
Serger (Pfaff)	140
Clothes Dryer (propane)	300
Clothes Iron	1200
Hair Curling Iron	55
Hair Dryer (high/low)	1,500/400
Furnace Fan (1/3 hp / 1/2 hp)	700/875
Guitar Amp (ave. volume)	45
Jimi Hendrix volume	8,500
Takagi T-K3 Water Heater, at rest	7
when heating water	34

Appliance	Watt Hours
Clothes Washer (front-loading)	145 watt-hours/load
Air Conditioning	1,500 watts/ton (or 12,000 Btu/hour) of capacity

*We have not listed refrigerators, freezers or dishwashers since their efficiency is getting better every year. Look at the **EnergyStar.gov** website for the latest ratings for these appliances and more.*

U.S. Annual Wind Average

To view this map and individual states in color:
http://rredc.nrel.gov/wind/pubs/atlas/maps.html

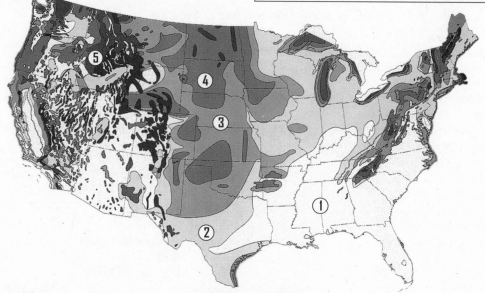

Wind Power Class	10 m (33 ft) tower		50 m (164 ft) tower	
	Wind Power W/m²	Speed m/s (mph)	Wind Power W/m²	Speed m/s (mph)
	0 — 0		0 — 0	
1	100 — 4.4 (9.8 mph)		200 — 5.6 (12.5 mph)	
2	150 — 5.1 (11.5 mph)		300 — 6.4 (14.3 mph)	
3	200 — 5.6 (12.5 mph)		400 — 7.0 (15.7 mph)	
4	250 — 6.0 (13.4 mph)		500 — 7.5 (16.8 mph)	
5	300 — 6.4 (14.3 mph)		600 — 8.0 (17.9 mph)	
6	400 — 7.0 (15.7 mph)		800 — 8.8 (19.7 mph)	
7	1000 — 9.4 (21.1 mph)		2000 — 11.9 (26.6 mph)	

System Sizing Worksheet — Your Electrical Needs

Electrical Device	Wattage (volts x amps)	X	Hours of Daily Use	X	Days Used per Week	÷	7	=	Ave. Daily Watt-Hours
		X		X		÷	7		
		X		X		÷	7		
		X		X		÷	7		
		X		X		÷	7		
		X		X		÷	7		
		X		X		÷	7		
		X		X		÷	7		
		X		X		÷	7		
		X		X		÷	7		
		X		X		÷	7		
		X		X		÷	7		
		X		X		÷	7		
		X		X		÷	7		
		X		X		÷	7		
		X		X		÷	7		

Total Average Watt-Hours per Day

To measure energy usage of electrical appliances, use a Watts-Up meter, or a Kill A Watt meter.
Actual usage will often vary greatly from the listed rating.

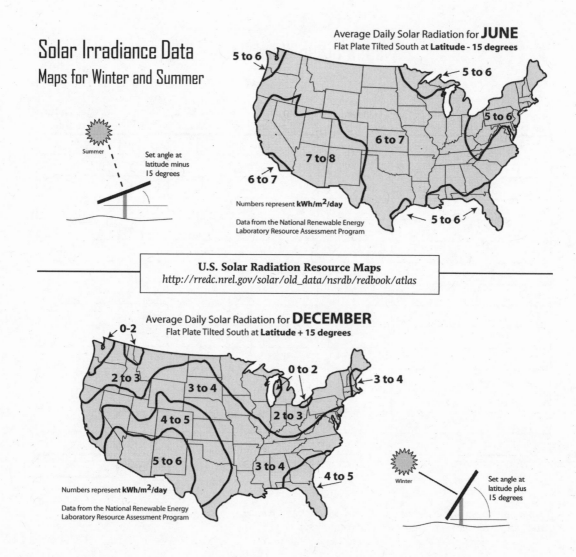

Solar Irradiance Data
Maps for Winter and Summer

Summer

Set angle at
latitude minus
15 degrees

Average Daily Solar Radiation for JUNE
Flat Plate Tilted South at **Latitude - 15 degrees**

5 to 6

5 to 6

5 to 6

6 to 7

7 to 8

6 to 7

5 to 6

Numbers represent **kWh/m^2/day**

Data from the National Renewable Energy
Laboratory Resource Assessment Program

U.S. Solar Radiation Resource Maps
http://rredc.nrel.gov/solar/old_data/nsrdb/redbook/atlas

Average Daily Solar Radiation for DECEMBER
Flat Plate Tilted South at **Latitude + 15 degrees**

0-2

0 to 2

2 to 3

3 to 4

3 to 4

4 to 5

2 to 3

5 to 6

3 to 4

4 to 5

Numbers represent **kWh/m^2/day**

Data from the National Renewable Energy
Laboratory Resource Assessment Program

Winter

Set angle at
latitude plus
15 degrees

Solar Array Sizing Worksheet

	June	Example	December
1. Input your **Average Watt-Hours per Day** from page 171.	_____	3000	_____
2. Find your site on the maps on the opposite page and input the nearest figure.	_____	6	_____
3. To find the number of watts you need to generate per hour of full sun, divide line 1 by line 2.	_____	500	_____
4. Select a solar module and multiply its rated wattage by .70 (.80 if using an MPPT charge controller). *Example: Enter 84 for a 120-watt module (or 96 with MPPT).**	_____	84	_____
5. To find the number of modules needed, divide line 3 by line 4. **	_____	**6**	_____

* You'll only get 120 watts from a 120-watt PV module when using an MPPT charge controller during the 2 hours nearest high noon, and **only** when the surface temperature of the module is below 77°F ... which is hardly ever. Output is typically derated to 60% - 70% for standard charge controllers (75% - 80% with MPPT) to give you a more accurate number.

** The exact number of modules you need will be determined by the system voltage, since modules are wired together in series strings to achieve a voltage within the range required by the charge controller or inverter.

State Energy Offices

*The ✗ symbol denotes a state that does **not** support net metering at the state level (as of early 2009).*

Many states also have utilities that support net metering (with and without a statewide law). More information can be found at: **www.dsireusa.org**

Alabama Energy Office ✗
Dept. of Economic and
  Community Affairs
401 Adams Avenue
P.O. Box 5690
Montgomery, AL 36103-5690
Call: (334) 242-5100
Fax: (334) 242-5099
www.adeca.alabama.gov

Alaska Energy Authority ✗
Alaska Industrial Development
  and Export Authority
813 W. Northern Lights Blvd.
Anchorage, AK 99503
Call: (907) 771-3000
Fax: (907) 771-3044
www.aidea.org/aea

Arizona Energy Office
Arizona Dept. of Commerce
1700 W. Washington Street,
  Suite 220
Phoenix, AZ 85007
Call: (602) 771-1194
Fax: (602) 771-1203
www.azcommerce.com/Energy/

Arkansas Energy Office
Arkansas Economic Development
  Commission
One Capitol Mall
Little Rock, AR 72201
Call: (501) 682-1121
Fax: (501) 682-7394
www.1800arkansas.com/Energy/

California Energy Commission
Renewable Energy Program
1516 Ninth Street, MS-29
Sacramento, CA 95814-5512
Call: (916) 654-4058
Toll Free in CA (800) 555-7794
Fax: (916) 654-4420
www.energy.ca.gov/renewables/

Colorado Energy Office
Governor's Energy Office
1580 Logan Street, Suite 100
Denver, CO 80203
Call: (303) 866-2100
Toll Free: (800) 632-6662
Fax: (303) 866-2930
www.colorado.gov/energy/

Connecticut Energy Office
Connecticut Office of Policy and
  Management
450 Capitol Avenue
Hartford, CT 06106-1379
Call: (860) 418-6200
Toll Free: (800) 286-2214
Fax: (860) 418-6496
www.ct.gov/opm

Delaware Energy Office
Department of Natural Resources
  and Environmental Control
1203 College Park Drive, Ste. 101
Dover, DE 19904
Call: (302) 735-3480
Fax: (302) 739-1840
www.delaware-energy.com

District of Columbia Energy Office
District Dept. of the Environment
51 N Street NE
Washington, DC 20002
Call: (202) 535-2600
Fax: (202) 535-2881
http://ddoe.dc.gov/

Florida Energy Office
Governors' Office of Energy &
  Climate Change
600 South Calhoun St., Ste. 251
Tallahassee, FL 32399-1300
Call: (850) 487-3800
Fax: (850) 922-9701
www.dep.state.fl.us/energy/

Georgia Division of Energy Resources
Georgia Environmental Facilities
  Authority
233 Peachtree Street NE
Harris Tower, Suite 900
Atlanta, GA 30303
Call: (404) 584-1000
Fax: (404) 584-1069
www.gefa.org/

Hawaii Energy Office
Dept. of Business, Economic
  Development & Tourism
235 S. Beretania St., Room 502
P.O. Box 2359
Honolulu, Hawaii 96804-2359
Call: (808) 587-3807
Fax: (808) 586-2536
www.hawaii.gov/dbedt/info/energy

**Idaho Office of Energy
Resources** ✗
322 East Front Street
P.O. Box 83720
Boise, Idaho 83720
Call: (208) 287-4891
Fax: (208) 287-6713
www.energy.idaho.gov

**Illinois Energy Bureau &
Recycling**
Illinois Dept. of Commerce and
  Economic Opportunity
620 East Adams Street
Springfield, IL 62701
Call: (217) 785-3416
Fax: (217) 785-2618
*www.ilbiz.biz/dceo/Bureaus/
  Energy_Recycling/*

**Indiana Office of Energy
Development**
101 W. Ohio Street, Suite 1250
Indianapolis, IN 46204
Call: (317) 232-8939
Fax: (317) 232-8995
www.in.gov/oed/

**Iowa Office of Energy
Independence**
Lucas State Office Building
321 East 12th Street
Des Moines, IA 50319
Call: (515) 281-0187
Fax: (515) 281-4225
www.energy.iowa.gov

Kansas State Energy Office ✗
Kansas Corporation Commission
1300 SW Arrowhead Road
Topeka, Kansas 66604-4074
Call: (785) 271-3170
Fax: (785) 271-3268
www.kcc.state.ks.us/energy/

**Kentucky Dept. for Energy
Development and
Independence**
500 Mero Street, 12th Floor,
  Capital Plaza Tower
Frankfort, KY 40601
Call: (502) 564-7192
Toll-free in KY: (800) 282-0868
Fax: (502) 564-7484
www.energy.ky.gov

Louisiana Energy Office
Technology Assessment Division
Dept. of Natural Resources
P.O. Box 94396
617 North Third Street
Baton Rouge, LA 70804-9396
Call: (225) 342-1399
Fax: (225) 342-1397
*http://dnr.louisiana.gov/sec/
  execdiv/techasmt/*

Efficiency Maine
Maine Public Utilities
  Commission
242 State Street
Augusta, ME 04333-0018
Call: (866) 376-2463
Fax: (207) 287-1039
www.efficiencymaine.com

Maryland Energy Office
Maryland Energy Administration
1623 Forest Drive, Suite 300
Annapolis, MD 21403
Call: (410) 260-7655
Toll Free: (800) 72-ENERGY
Fax: (410) 974-2250
www.energy.maryland.gov

**Massachusetts Department of
Energy Resources**
Executive Office of Energy and
  Environmental Affairs
100 Cambridge Street, Suite 1020
Boston, MA 02114
Call: (617) 626-7300
Fax: (617) 727-0030
www.magnet.state.ma.us/doer

Michigan Energy Office
Department of Energy, Labor &
  Economic Growth
611 W Ottawa
P.O. Box 30221
Lansing, MI 48909
Call: (517) 241-6228
Fax: (517) 241-6229
www.michigan.gov/dleg
  (Look under *Inside DELEG /
  Energy Office*)

**Minnesota Office of Energy
Security**
Minnesota Dept. of Commerce
85 7th Place East, Suite 500
St. Paul, MN 55101
Call: (651) 296-5175
Toll free in MN: (800) 657-3710
Fax: (651) 297-7891
www.energy.mn.gov

Mississippi Energy Division ✗
MS Development Authority
501 North West Street
P.O. Box 849
Jackson, MS 39205
Call: (601) 359-6600
Toll free: (800) 222-8311
Fax: (601) 359-6642
www.mississippi.org
  (Look under *Energy*)

Missouri Energy Center
Dept. of Natural Resources
P.O. Box 176
Jefferson City, MO 65102-0176
Call: (573) 751-3443
Toll free: (800) 361-4827
Fax: (573) 751-6860
www.dnr.mo.gov/energy/

Montana Energy Office
Dept. of Environmental Quality
P.O. Box 200901
1100 North Last Chance Gulch
Helena, MT 59620-0901
Call: (406) 841-5240
Fax: (406) 841-5222
www.deq.mt.gov/energy/

Nebraska Energy Office ✗
P.O. Box 95085
1111 "O" Street, Suite 223
Lincoln, NE 68509-5085
Call: (402) 471-2867
Fax: (402) 471-3064
www.neo.ne.gov

Nevada State Office of Energy
Office of the Governor
727 Fairview Drive, Suite F
Carson City, NV 89701
Call: (775) 687-9700
Fax: (775) 687-9714
www.energy.state.nv.us

**New Hampshire Office of
Energy and Planning**
4 Chenell Dive
Concord, NH 03301-8501
Call: (603) 271-2155
Fax: (603) 271-2615
www.nh.gov/oep

**New Jersey's Clean Energy
Program**
NJ Board of Public Utilities
P.O. Box 350
44 S. Clinton Avenue
Trenton, NJ 08625-0350
Call: (609) 777-3335
Fax: (609) 777-3330
www.njcleanenergy.com

**New Mexico Energy
Conservation and
Management Division**
New Mexico Energy, Minerals
   and Natural Resources Dept.
1220 S. St. Francis Drive
P.O. Box 6429
Santa Fe, NM 87505
Call: (505) 476-3310
Fax: (505) 476-3322
www.emnrd.state.nm.us/ecmd

**New York State Energy
Research and Development
Authority**
17 Columbia Circle
Albany, NY 12203-6399
Call: (518) 862-1090
Toll free: (866) NYSERDA
Fax: (518) 862-1091
www.nyserda.org

**North Carolina State Energy
Office**
NC Dept. of Administration
1830-A Tillery Place
1340 Mail Service Center
Raleigh, NC 27699-1340
Call: (919) 733-2230
Toll free in NC: (800) 662-7131
Fax: (919) 733-2953
www.energync.net

**North Dakota Office of
Renewable Energy & Energy
Efficiency**
Department of Commerce
Division of Community Services
P.O. Box 2057
1600 E. Century Avenue, Suite 2
Bismarck, ND 58502-2057
Call: (701) 328-5300
Fax: (701) 328-2308
*www.communityservices.nd.gov/
   energy/*

Ohio Energy Office
Ohio Dept. of Development
77 South High Street
P.O. Box 1001
Columbus, OH 43216-1001
Call: (614) 466-6797
Fax: (614) 466-1864
www.odod.state.oh.us/cdd/oee/

Oklahoma State Energy Office
Oklahoma Dept. of Commerce
900 N. Stiles
P.O. Box 26980
Oklahoma City, OK 73126-0986
Call: (405) 815-6552
Toll free: (800) 879-6552
Fax: (405) 605-2807
www.okcommerce.gov (Look
   under *Communities* for *State
   Energy Office*)

Oregon Department of Energy
625 Marion Street, NE
Salem, OR 97301-3737
Call: (503) 378-4040
Toll free: (800) 221-8035
Fax: (503) 373-7806
www.oregon.gov/energy

Pennsylvania Office of Energy & Technology Deployment
Department of Environmental
    Protection
Rachel Carson State Office Bldg.
400 Market Streest
P.O. Box 8772
Harrisburg, PA 17105-8772
Call: (717) 783-0540
Fax: (717) 783-0546
www.depweb.state.pa.us/energy/

Rhode Island Office of Energy Resources
One Capital Hill
Providence, RI 02908
Call: (401) 574-9100
Fax: (401) 574-9125
www.energy.ri.gov

South Carolina Energy Office ✗
SC Budget and Control Board
1200 Senate Street
408 Wade Hampton Building
Columbia, SC 29201
Call: (803) 737-8030
Toll free: (800) 851-8899
Fax: (803) 737-9846
www.energy.sc.gov

South Dakota Energy Management Office ✗
Bureau of Administration
523 E. Capitol Avenue
Pierre, SD 57501-3182
Call: (605) 773-3899
Fax: (605) 773-5980
www.state.sd.us/boa/ose/
    OSE_Statewide_Energy.htm

Tennessee Energy Division ✗
Department of Economic &
    Community Development
312 Rosa L. Parks Ave., 10th Floor
Nashville, Tennessee 37243-1102
Call: (615) 741-2994
Fax: (615) 741-0607
www.state.tn.us/ecd/energy.htm

Texas Energy Office ✗
State Energy Conservation Office
111 E. 17th Street, #1114
Austin, TX 78701
Call: (512) 463-1931
Fax: (512) 475-2569
www.seco.cpa.state.tx.us

Utah State Energy Program
1594 W. North Temple, Ste. 3110
P.O. Box 146100
Salt Lake City, UT 84114-6100
Call: (801) 537-3300
Fax: (801) 537-3400
http://geology.utah.gov/sep/

Vermont Energy Efficiency, Conservation and Renewable Energy
112 State Street, Drawer 20
Montpelier, VT 05620-2601
Call: (802) 828-2811
Toll free: (800) 622-4496
Fax: (802) 828-2342
www.publicservice.vermont.gov/
    energy-efficiency/energy-
    efficiency.html

Virginia Division of Energy
Washington Building, 8th Floor
1100 Bank Street
Richmond, VA 23219
Call: (804) 692-3218
Fax: (804) 692-3238
www.dmme.virginia.gov/
    divisionenergy.shtml

Washington State Energy Policy Division
Dept. of Community, Trade &
    Economic Development
906 Columbia St. SW
P.O. Box 43173
Olympia, WA 98504-3173
Call: (360) 725-3118
Fax: (360) 586-0049
www.cted.wa.gov (Look under
    Energy Policy)

West Virginia Division of Energy
West Virginia Dept. of Commerce
Capitol Complex, Bldg 6, Rm 553
1900 Kanawha Blvd. E.
Charleston, WV 25305-0311
Call: (304) 558-2234
Toll free: (800) 982-3386
Fax: (304) 558-0449
www.energywv.org/community/
    eep.html

Wisconsin Office of Energy Independence
17 West Main Street, Ste. 429
Madison, WI 53703
Call: (608) 261-6609
Fax: (608) 261-8427
www.energyindependence.wi.gov

Wyoming State Energy Program
Wyoming Business Council
214 West 15th Street
Cheyenne, WY 82002-0240
Call: (307) 777-2800
Toll free: (800) 262-3425
Fax: (307) 777-2838
www.wyomingbusiness.org/
    business/energy.aspx

Puerto Rico Energy Office
Department of Natural &
    Environmental Resources
PO Box 9066600
Puerta de Tierra Station
San Juan, Puerto Rico 00906
Call: (787) 724-8774 ext. 4015
Fax: (787) 721-3089

*The ✗ symbol denotes a state that does **not** support net metering at the state level (as of early 2009). Current information can be found at: **www.dsireusa.org***

Resources

Additional resources can be found on the internet.

PV MANUFACTURERS

BP Solar
www.bpsolar.com

Evergreen Solar
www.evergreensolar.com

GE Solar
www.gepower.com/prod_serv/
products/solar/en/index.htm

Kyocera Solar
www.kyocerasolar.com

Mitsubishi Electric
http://global.mitsubishi
electric.com/bu/solar/

Sharp Solar Systems
http://solar.sharpusa.com

SolarWorld Industries
(formerly Shell Solar)
www.solarworld-usa.com

SunPower Corporation
http://us.sunpowercorp.com

Uni-Solar® Products
www.uni-solar.com

MOUNTING HARDWARE MANUFACTURERS

Direct Power and Water
www.directpower.com

Next Gen Energy
www.ngeus.com

Professional Solar Products
www.prosolar.com

Unirac, Inc.
www.unirac.com

INVERTER MANUFACTURERS

Beacon Power
www.beaconpower.com

Enphase Energy
www.enphaseenergy.com

Fronius USA Solar Electronics
www.fronius.com

OutBack Power Systems
www.outbackpower.com

PV Powered
www.pvpowered.com

SMA
www.sma-america.com

Solectria
www.solren.com

Xantrex Technology
www.xantrex.com

SOLAR HEATING

Takagi
www.takagi.com

TCT Solar
www.tctsolar.com

Themo Technologies
www.thermomax.com

SolaHart
www.solahart.com

Your Solar Home
www.yoursolarhome.com

WIND RESOURCES

Abundant Renewable Energy
www.abundantre.com

Bergey Wind Power
www.bergey.com

Kestrel
www.kestrelwind.co.za

Proven Wind Turbines
www.solarwindworks.com

Southwest Windpower Inc.
www.windenergy.com

OTHER COMPANIES NOTED IN THIS BOOK

Affinity Energy
www.affinityenergy.com

Concorde Battery Corporation
www.concordebattery.com

DC Power Systems
www.dcpower-systems.com

Fat Spaniel
www.fatspaniel.com

groSolar
www.grosolar.com

Hawker Batteries
www.hawkerpowersource.com

Mississippi Solar
www.MSSolar.net

MK Battery
www.mkbattery.com

Renewable Power Solutions
www.RPS-Solar.com

Right Hand Engineering
www.righthandeng.com

Standard Solar
www.standardsolar.com

Summit Electrical Service
www.summit-e.com

Sunlight Solar Energy
www.sunlightsolar.com

Triangle Electrical Systems
www.trianglesystems.com

EDUCATION & CLASSES

Institute for Solar Living
www.solarliving.org

Midwest Renewable Energy Association
www.the-mrea.org

Solar Energy International
www.solarenergy.org

ORGANIZATIONS & REFERENCE WEBSITES

American Solar Energy Society
www.ases.org

American Wind Energy Assn.
www.awea.org

Center for Renewable Energy & Sustainable Technology
www.crest.org

Contractors License Reference Site
www.contractors-license.org

Database of State Incentives for Renewable Energy
www.dsireusa.org

Efficient Windows Collaborative
www.efficientwindows.org

Energy Star (*energy ratings*)
www.energystar.gov

Find Solar
www.findsolar.com

Florida Solar Energy Center
www.fsec.ucf.edu

Geothermal Heat Pump Consortium
www.geoexchange.org

International Solar Energy Society
www.ises.org

Interstate Renewable Energy Council
www.irecusa.org

NAHB National Green Building Program
www.NAHBgreen.org

North American Board of Certified Energy Practitioners (NABCEP)
www.nabcep.org

Renewable Energy Access
www.renewableenergy access.com

Rocky Mountain Institute
www.rmi.org

SolarBuzz
www.solarbuzz.com

SRCC (Solar Rating and Certification Corporation)
www.solar-rating.org

Solar Energy Industries Assn.
www.seia.org

U.S. Department of Energy's EERE (Energy Efficiency and Renewable Energy)
www.eere.energy.gov

U.S. Green Building Council
www.usgbc.org

U.S. Solar Radiation Maps
http://rredc.nrel.gov/solar/ old_data/nsrdb/redbook/atlas

Wind Energy Maps & Tables
http://rredc.nrel.gov/wind/ pubs/atlas/maps.html

http://rredc.nrel.gov/wind/ pubs/atlas/tables.html

MAGAZINES

***BackHome* Magazine**
www.backhomemagazine.com

Home Power
www.homepower.com

Mother Earth News
www.motherearthnews.com

Solar Pro
www.solarprofessional.com

Solar Today
www.solartoday.org

More Books and Reference Reading

Consumer Guide to Home Energy Savings; American Council for an Energy Efficient Economy. An abundance of tips & proven advice with lists of top energy-efficient appliances.

A Guide to Photovoltaic (PV) System Design and Installation; Endecon Engineering and Regional Economic Research for the California Energy Commission. This guide (PDF file) to properly installing grid-tie systems has been the model for many California installers. *www.energy.ca.gov/reports/2001-09-04_500-01-020.PDF*

Homebrew Wind Power: A Hands-On Guide to Harnessing the Wind by Dan Bartmann and Dan Fink; Buckville Publications. For anyone wishing to build their own wind turbine for a battery-based system, this book can't be beat. Step-by-step instructions are buttressed by clear explanations of the fundamental principles involved in turbine operations.

Natural Home Heating: The Complete Guide to Renewable Energy Options by Greg Pahl; Chelsea Green Publishing. A well-organized tour of renewable home-heating options, including wood, pellet, corn and grain-fired stoves, fireplaces, furnaces, boilers, masonry heaters, active and passive solar systems, and heat pumps.

Photovoltaic Systems by Jim Dunlop and NJATC. A comprehensive textbook (and CD-Rom) to the design, installation, and evaluation of residential and commercial photovoltaic (PV) systems.

Photovoltaics: Design and Installation Manual, by Solar Energy International. A complete and up-to-date PV book available for pros or seriously interested homeowners.

Photovoltaic Power Systems and the National Electrical Code: Suggested Practices; J. Wiles, Southwest Technology Development Institute. A plain-english explanation of PV-related electric code. Updated and republished periodically.

Power with Nature 2nd edition: Alternative Energy Solutions for Homeowners by Rex Ewing; PixyJack Press. Focused on off-grid renewable energy (solar, wind, micro-hydro), plus home heating, water pumping and more.

The Solar House: Passive Heating and Cooling by Dan Chiras; Chelsea Green Publishing. An up-to-date primer on regional passive solar design—the only intelligent way to build a house.

Solar Living Sourcebook; Real Goods. A comprehensive retail guide to renewable energy. Part textbook, part catalog. Also contains J. Wiles Suggested Practices (listed above).

Solar Water Heating: A Comprehensive Guide to Solar Water and Space Heating Systems by Bob Ramlow with Benjamin Nusz; Mother Earth News Wiser Living Series. A great source for in-depth information on solar water heating.

Wind Energy Basics by Paul Gipe; Chelsea Green Publishing. A guide to small and micro wind systems. All the experience, advice, and resources of Paul's complete book (*Wind Power: Renewable Energy for Home, Farm and Business*), but without the big commercial and industrial turbines.

Glossary

Absorption Stage A stage of the battery-charging process performed by the charge controller, where the batteries are held at the bulk-charging voltage for a specified time period, usually one to two hours.

Alternating Current (AC) Electric current that reverses its direction of flow at regular intervals, usually many times per second; common household current is AC.

Alternative Energy See Renewable Energy.

Amorphous Solar Cell Type of solar cell constructed by using several thin layers of molten silicon. Amorphous solar cells cost less to produce and perform better in sub-optimal lighting conditions, but need more surface area than conventional crystalline cells to produce an equal amount of power.

Ampere (Amp) Unit of electrical current, thus the rate of electron flow. One volt across one ohm of resistance is equal to a current flow of one ampere.

Ampere Hour (AH) A current of one ampere flowing for one hour. Used primarily to rate battery capacity and solar or wind output.

Anti-Freeze See Glycol.

Array See Photovoltaic Array.

Batch Collector An open-loop solar hot-water collector designed to continuously hold and heat several gallons of water. They do not require heat exchangers since they are plumbed directly into the home's domestic hot water line. Although simple and inexpensive, they are only practical in non-freezing climates.

Battery Electrochemical cells enclosed within a single container and electrically interconnected in a series / parallel arrangement designed to provide a specific DC operating voltage and current level. Batteries for PV systems are commonly 6- or 12-volts, and are used in 12, 24 or 48-volt operations.

Battery Cell The basic functional unit in a storage battery. It consists of one or more positive electrodes or plates, an electrolyte that permits the passage of charged ions, one or more negative electrodes or plates, and the separators between plates of opposite polarity.

Battery Capacity Total amount of electrical current, expressed in ampere-hours (AH), that a battery can deliver to a load under a specific set of conditions.

Battery Life Period during which a battery is capable of operating at or above its specified capacity or efficiency level. A battery's useful life is generally considered to be over when a fully charged cell can only deliver 80% of its rated capacity.

Btu (British thermal unit) The amount of heat energy required to raise one pound of water one degree Fahrenheit at one atmosphere pressure. In the U.S. it is the common unit of measure for rating water and air heating systems.

Building-integrated PV (BIPV) Where PV is integrated into a building, replacing conventional materials, such as siding, shingles or roofing panels.

Bulk Stage Initial stage of battery charging, where the charge controller allows maximum charging in order to reach the bulk voltage setting.

Cell See Photovoltaic Cell.

Cell Efficiency Percentage of electrical energy that a solar cell produces (under optimal conditions) divided by the total amount of solar energy falling on the cell; typically 12% – 15%.

Charge Controller Component between the solar array or wind turbine and the battery bank that brings the batteries to an optimal state of charge, without over-charging them. MPPT charge controllers go a step further, by converting excess array voltage into usable amperage.

Circuit A system of conductors connected together for the pur-pose of carrying an electric cur-rent from a generating source, through the devices that use the electricity (the loads), and back to the source.

Circuit Breaker Safety device that shuts off power (i.e. it cre-ates an open circuit) when it senses too much current.

Closed-Loop System When applied to solar or geothermal heating systems, a system in which the heat-collection fluid is kept isolated from the domestic water supply and/or the environ-ment. A heat-exchanger is there-fore required to transfer heat from the heat source to the heating medium (water or air).

Collector *See* solar collector.

Convection Natural currents flowing in air and water, resulting from the fact that hot air (or water) rises. In a passively cooled home, warm air rises to a high exit point in the structure, drawing cool air from a low-lying shaded entry point (such as a north-facing basement window).

Conversion Efficiency *See* Cell Efficiency.

Current Flow of electricity between two points. Measured in amps.

Depth of Discharge (DOD) The ampere-hours removed from a fully charged battery, expressed as a percentage of rated capacity. For optimum health in most bat-teries, the DOD should never exceed 50%.

Direct Current (DC) Electrical current that flows in only one direction. It is the type of current produced by solar cells, and the only current that can be stored in a battery.

Distributed System A system installed near where the elec-tricity is used, as opposed to a central system—such as a coal or nuclear power plant—that sup-plies electricity to the electrical grid. A grid-tied residential solar system is a distributed system.

Electrical Grid A large distribu-tion network—including towers, poles, and transmission lines—that delivers electricity over a wide area.

Electric Circuit *See* Circuit.

Electric Current *See* Current.

Electricity In a practical sense, the controlled flow of electrons through a conductor. In a scien-tific sense, the non-gravitational and non-nuclear repulsive and attractive forces governing much of the behavior of charged sub-atomic particles.

Electrode A conductor used to lead current into or out of a nonmetallic part of a circuit, such as a battery's positive and negative electrodes.

Electrolyte Fluid used in bat-teries as the transport medium for positively and negatively charged ions. In lead-acid batteries, it is a somewhat diluted sulfuric acid.

Electron Negatively charged particle. An electrical current is a stream of electrons moving through an electrical conductor.

Emittance A measure of the ability of a material to emit radiant energy. *See also* Low-E Windows.

Emissivity *See* Emittance.

Energy The capacity for per-forming work. Solar cells convert electromagnetic energy (light) from the sun into electrical energy, while wind turbines con-vert the kinetic energy of the air into first mechanical energy, and then electrical energy.

Energy Audit An inspection process that determines how much energy you use in your home, usually accompanied by specific suggestions for saving energy.

Equalization A controlled process of overcharging non-sealed lead-acid batteries, intended to clean lead sulfates from the battery's plates, and restore all cells to an equal state-of-charge.

Evacuated Tube (ET) Each tube consists of two nested glass tubes separated by a vacuum. Within the inner tube a copper heat pipe transfers solar energy to a fluid which, in turn, gives up its heat to a manifold. The manifold holds several tubes to form a evacuated tube collector.

Flat Plate Collector A popular type of solar hot-water panel, constructed as a shallow, glazed, aluminum box containing a serpentine loop of black copper piping through which runs a heat-collection fluid, such as water or propylene glycol.

Float Stage A battery-charging operation performed by the charge controller in which enough energy is supplied to meet all loads, plus internal component losses, thus always keeping the battery up to full power and ready for service.

Fossil Fuels Carbon- and hydrogen-laden fuels formed underground from the remains of long-dead plants and animals. Crude oil, natural gas and coal are fossil fuels.

Full Sun Scientific definition of solar power density received at the surface of the earth at noon on a clear day. Defined as 1,000 watts per square meter (W/m^2). Reality varies from 600 to 1,200 W/m^2, depending on latitude, altitude, and atmospheric purity.

Geoexchange *See* Geothermal.

Geothermal Heat energy from the earth. Geothermal heat pumps used in space heating/cooling systems utilize this energy by moving heat to and from the earth, which is an almost constant temperature just a few feet underground.

Glazing Single or multiple layers of glass, as a windowpane or a covering for a solar collector.

Glycol Antifreeze; propylene glycol is a non-toxic liquid used to prevent freeze damage to outdoor-mounted piping and components in solar water-heating systems.

Greenhouse Effect A warming effect that occurs when heat from the sun becomes trapped in the Earth's atmosphere due to the heat-absorbing properties of certain (greenhouse) gases.

Greenhouse Gases Gases responsible for trapping heat from the sun within the Earth's atmosphere. Water vapor and carbon dioxide are the most prevalent, but methane, ozone, chlorofluorocarbons and nitrogen oxides are also greenhouse gases.

Grid *See* Electrical Grid.

Grid-Tie PV System Solar PV system that is tied into (connected to) the utility's electrical grid. When generating more power than necessary to power all its loads, the grid-tie home system sends the surplus to the grid. At night, the system draws power from the grid.

Ground-Source Heat Pump In home heating systems, a type of heat pump that extracts heat from the earth, either the from the soil or from ground water.

Heat Exchanger Component in solar water heating and geothermal systems that transfers heat from one medium to another without allowing the two to mix.

Hertz (Hz) A unit denoting the frequency of an electromagnetic wave, equal to one cycle per second. In alternating current, the frequency at which the current switches direction. In the U.S. this is usually 60 cycles per second (60 Hz).

Hybrid System Power-generating system consisting of two or more subsystems, such as a wind turbine or diesel generator, and a photovoltaic system.

Insolation *See* Irradiance.

Inverter Component that transforms direct current (DC) flowing from a solar system or battery to alternating current (AC) for use in the home.

Irradiance Rate at which radiant energy arrives at a specific area of the Earth's surface during a specific time interval. Measured in W/m^2.

Junction Box (J-Box) Enclosure on the back of a solar module where it is connected (wired) to other solar modules.

Kilowatt (kW) Unit of electrical power, equal to one thousand watts.

Kilowatt-Hour (kWh) One thousand watts being used over a period of one hour. The kWh is the usual billing unit of energy for utility companies.

Life-Cycle Cost Estimated cost of owning, operating, and disposing of a system over its useful life.

Load Anything that draws power from an electrical circuit.

Low-E Windows Windows that have been coated with substances—usually metal oxides—that block the flow of heat (emittance or emissivity) but permit light to pass unimpeded. It's effectiveness is measured by the U-factor.

Maximum Power Point Tracking (MPPT) Technology used by some inverters and charge controllers to convert, through the use of DC-DC power converters, excess array voltage into usable amperage, by tracking the optimal power point of the I-V curve.

Megawatt (MW) One million watts; 1,000 kilowatts. Utility power plants and wind farms are usually rated in megawatts.

Module *See* photovoltaic module.

Monocrystalline Solar Cell Type of solar cell made from a thin slice of a single large silicon crystal. Also known as single-crystal solar cell.

Multicrystalline Solar Cell *See* polycrystalline solar cell.

National Electrical Code (NEC) The U.S. minimum inspection requirements for all types of electrical installations, including solar/wind systems.

NEMA (National Electrical Manufacturers Association) The U.S. trade organization which sets standards for the electrical manufacturing industry.

NREL (National Renewable Energy Laboratory) Based in Golden, Colorado, NREL is the principal research laboratory for the Department of Energy's Office of Energy Efficiency and Renewable Energy. They study, test and develop cutting-edge renewable energy technologies.

Net Metering A practice used in conjunction with a solar- or wind-electric system. The electric utility's meter tracks the home's net power usage, spinning forward when electricity is drawn from the utility, and spinning backward when the solar or wind system is generating more electricity than is currently needed to run the home's loads.

Ohm Measure of the resistance to current flow in electrical circuits, equal to the amount of resistance overcome by one volt in causing one ampere to flow.

Oh, Wow! Exclamatory colloquialism frequently uttered by neophyte solar enthusiasts upon seeing their electric meters spin backward for the first time.

On-Demand Water Heaters *See* Tankless Heaters.

Orientation Describes the direction that a solar array or collector faces. The two components of orientation are the tilt angle (the number of degrees the panel is raised from the horizontal position) and the aspect angle, (the degree by which the panel deviates from facing due south).

Panel *See* Solar Panel.

Parallel Connection Wiring configuration whereby the current is given more than one path to follow, thus amperage is increased while voltage remains unchanged. In DC systems, parallel wiring is positive to positive (+ to +) and negative to negative (- to -). *See also* Series Connection.

Passive Solar Home Home designed to use sunlight for direct heating and lighting, without circulating pumps or energy conversion systems. This is achieved through the use of energy efficient materials (such as windows, skylights and Trombe walls) and proper design and orientation of the home.

Peak Load Maximum amount of electricity being used at any one point during the day.

Photon Basic unit of light. A photon can act as either a particle or a wave, depending on how it's activity is measured. The shorter the wavelength of a stream of photons, the more energy these photons possess. This is why ultraviolet (UV) light is so destructive, while infrared (IR) is not.

Photovoltaic (PV) Refers to the technology of converting sunlight directly into electricity, through the use of photovoltaic (solar) cells.

Photovoltaic Array A system of interconnected PV modules (solar panels) acting together to produce a single electrical output.

Photovoltaic Cell The basic unit of a PV (solar) module. Crystalline photovoltaic cells produce an electrical potential of around 0.5 volts. The higher voltages typical in PV modules are achieved by connecting solar cells together in series.

Photovoltaic Module Collection of solar cells joined as a unit within a single frame, commonly called a solar panel.

Photovoltaic System Complete set of interconnected components—including a solar array, inverter, etc.—designed to convert sunlight into usable electricity.

Polycrystalline Solar Cell Type of solar cell made from many small silicon crystals (crystallites). Also known as a multicrystalline solar cell.

PV Photovoltaic.

Rated Power Nominal power output of a RE component such as a wind turbine or inverter; some components cannot produce rated power continuously.

Renewable Energy (RE) Energy obtained from sources that are essentially inexhaustible (unlike, for example, fossil fuels, of which there is a finite supply). Renewable sources of energy include conventional hydroelectric power, wood, waste, geothermal, wind, photovoltaic, and solar-thermal energy.

Semiconductor Material that has an electrical conductivity between that of a metal and an insulator. Typical semiconductors for PV cells include silicon, gallium arsenide, copper indium diselenide, and cadmium telluride.

Series Connection A wiring configuration where the current is given but one path to follow, thus increasing voltage without changing the amperage. Series wiring is positive to negative (+ to -) or negative to positive (- to +). *See also* Parallel Connection.

Silicon (Si) The most common semiconductor material used in the manufacture of PV cells.

Single-Crystal Silicon *See* Monocrystalline Solar Cell.

Solar Air Panel A type of solar collector that heats and recirculates air from a room or house. The panel(s) can mounted either on the roof or on a south-facing outside wall.

Solar Cell *See* Photovoltaic Cell.

Solar Collector A component of a solar heating system that absorbs solar radiation to heat air or water. Most commonly used for domestic hot-water or space-heating systems.

Solar Energy Energy from the sun measured in watts per square meter (W/m^2). Virtually all energy on Earth—including solar, wind, hydroelectric and even fossil-fuel energy—originated as solar energy.

Solar Insolation *See* Irradiance.

Solar Module *See* Photovoltaic Module.

Solar Panel Commonly refers to photovoltaic modules that are used to make electricity.

Solar Power *See* Solar Energy.

Solar Thermal Term applied to systems that capture the sun's energy and use it to heat water or air. Solar hot water and solar air heating are two examples of solar thermal applications.

Solar Water Heating A system in which water or a heat collection fluid heated in solar collectors is circulated through a heat exchanger to provide hot water for household use and/or home or pool heating.

SRCC The Solar Rating and Certification Corporation is a non-profit organization whose primary purpose is the development and implementation of certification programs and national rating standards for solar energy equipment.

Stand-Alone A solar-electric system that operates without connection to the utility grid or another source of electricity. Typically, unused energy is stored in a battery bank to provide power at night.

Tankless Heaters A type of water heater, either electric or gas, that heats water instantly as it flows en route to the tap. They are more efficient than standard tank-type heaters since they are not required to keep a large volume of water constantly warm.

Thermal Collectors *See* solar collector, and Solar Air Panel.

Thermal Mass The ability of a material to absorb and store heat. Stone, concrete, tile, logs and water-filled containers all have good thermal mass, as does the earth.

Thermosiphon A method of solar water heating based on natural convection. Water (in an open loop or closed loop) circulating through a flat-plate collector rises into a highly insulated tank directly above the collector. Being passive, thermosiphon systems work without the need of sensors or pumps.

Thin Film *See* Amorphous Solar Cell.

Tilt Angle The angle of inclination of a module measured from the horizontal. The most productive tilt angle is one in which the surface of the module is exactly perpendicular to sun's rays.

TOU (Time of Use) Billing A rate structure used by utilities that charges customers higher rates for electricity used during peak times of the day, and lower rates for off-peak usage.

Trombe Wall A south-facing wall—generally masonry—placed behind a window or pane of glass, designed to absorb heat during the day and emit it back into the room at night.

U-Factor A measure of a window's ability to conduct heat. It is the reciprocal of the R-factor (a measure of a substance's resistance to heat flow), so a lower U-factor means greater resistance to heat flow and a more energy efficient window.

Volt (V) A unit of electrical force. It is equal to the amount of electromotive force that will cause a steady current of one ampere to flow through a resistance of one ohm.

Watt (W) Unit of electrical power used to indicate the rate of energy produced or consumed by an electrical device. One ampere of current flowing at a potential of one volt produces one watt of power. Wind turbines and PV modules are often rated in watts.

Watt-hour (Wh) Unit of energy equal to one watt of power being used or produced for one hour.

Wind Energy The kinetic energy present in wind, measured in watts per square meter (W/m^2). Wind turbines convert the kinetic energy into mechanical energy through the use of propeller blades, which in turn drive an alternator to produce electricity.

Italic numbers represent photos, graphs or illustrations.

Index

Meet the Authors

Rex Ewing has lived blissfully off-grid with solar and wind energy since 1999 when he left the dusty plains of Colorado and headed for the Rockies to build his wife, LaVonne, a long-promised log home. When he's not writing books or magazine articles about renewable energy—or his first love, horses— he and LaVonne are probably trekking through the backcountry, canoeing, or enjoying the 50-mile view from their deck.

Before moving to the mountains to concentrate on his writing, Ewing raised grass hay and high-strung Thoroughbred race horses in the Platte River valley. Whenever his employees were clever enough to corral him behind a desk, he ran an equine nutrition firm, where he formulated and marketed a successful line of equine supplements worldwide.

Doug Pratt grew up in a wonderful rural community of Frank Lloyd Wright homes, but after twenty-three Michigan winters, he figured it wasn't going to get any better and ran off to northern California to be a good hippie and enjoy milder winters. Within a year the Arab Oil Embargo hit, which ignited a lifelong interest in renewable energy and energy conservation. He has lived in passive solar homes, on and off-grid since 1980, and worked in the RE industry as a technician, consultant, teacher, and writer since 1985. His house has been solar grid-tied since 2000 and his Toyota Prius was the first one sold into Mendocino County. And because the scenery is so beautiful, the roads so swoopy, and the climate so friendly in northern coastal California, his favorite recreation is two-up day trips on his BMW motorcycle.

also from **PIXYJACK PRESS**

Renewable Energy Titles

Power With Nature 2nd edition: Alternative Energy Solutions for Homeowners
REX A. EWING $25.00

HYDROGEN—Hot Stuff Cool Science: Discover the Future of Energy
REX A. EWING $25.00

Careers in Renewable Energy: Get a Green Energy Job
GREGORY MCNAMEE $20.00

Crafting Log Homes Solar Style: An Inspiring Guide to Self-Sufficiency
REX A. EWING & LAVONNE EWING $25.00

Visit our web site for a wealth of information about each title.

To order autographed copies:
PO Box 149 • Masonville, CO 80541
www.PixyJackPress.com

PIXYJACK PRESS INC

A 100% solar and wind-powered independent publisher since 1999.

AS A MEMBER OF THE GREEN PRESS INITIATIVE, WE PRINT OUR BOOKS ON CHLORINE-FREE, 100% RECYCLED PAPER.